CRUCIBLE 1972

The War for Peace in Vietnam

J. Keith Saliba

Battlespace Books
Jacksonville, Florida

Printed in the United States of America.

For more information, or to book an event, contact :

Email : ksaliba@ju.edu
Website : http://www.jkeithsaliba.com

ISBN - Paperback: 979-8-9871-7830-0

ISBN – Ebook - EPUB: 979-8-9871783-1-7

First Edition: January 2025

Dedicated to the 2.7 million Americans who served in Vietnam. Your efforts and sacrifice are remembered.

Acknowledgements

My sincere gratitude to each veteran of the 1972 war for peace who graciously shared his or her insights, or passed along documents, photographs, and other materials that helped enhance my understanding. In no particular order, I thank Chuck DeBellevue, Everett Alvarez, Bob Certain, Dick Francis, Jack Ensch, Mickey Mantel, John Oleson, Dean Vetter, Mark Tiedemann, Craig Honour, Charlie Plumb, Guy Gruters, Dick Evert, Jerry Tucker, Vince Massimini, Eddie Eiler, Jerry Phillippe, Steve Williams, and Tom Moe for sharing their experiences. I am also grateful to Steve Sherman, Andy Finlayson, Pat Reakes, Charlie Tupper, Mark Moyar, and the late, great Joe Galloway for their support and inspiration.

Glossary

ARVN – Army of the Republic of Vietnam (South Vietnam)

BDA – Bomb Damage Assessment

BUFF – Big Ugly Fat Fellow/Fucker (nickname for the B-52)

CIC – Combat Information Center

CINCSAC – Commander in Chief, Strategic Air Command

CMPC – Central Military Party Committee (North Vietnam)

CORDS – Civilian Operations and Revolutionary Development Support (joint U.S-RVN rural pacification program)

CSAR – Combat Search and Rescue

CVW – Carrier Air Wing (U.S. Navy)

DIA – Defense Intelligence Agency

DMZ – Demilitarized Zone (dividing North and South Vietnam)

DRV – Democratic Republic of Vietnam (North Vietnam)

ECM – Electronic Countermeasure

EOGB – Electro-Optically Guided Bomb

EWO – Electronic Warfare Officer

FAC – Forward Air Controller

FPJMC – Four-Party Joint Military Commission

GCI – Ground-Controlled Interception

GNP – Gross National Product

GO-GU – General Offensive-General Uprising

GVN – Government of Vietnam (South Vietnam)

H-Ks – Hunter-Killers

ICCS – International Commission of Control and Supervision (1973-1975)

ICSC – International Commission of Supervision and Control (1954-1973)

JGS – Joint General Staff (RVN)

KIA – Killed in Action

LGB – Laser Guided Bomb

MAG – Marine Aircraft Group

MIA – Missing in Action

MIGCAP – MiG Combat Air Patrol

NCA – National Command Authority

NCO – Noncommissioned Officer

NLF/VC – National Liberation Front/Viet Cong (South Vietnamese communist insurgency)

NCNRC – National Council of National Reconciliation and Concord

NSC – National Security Council

PACOM – U.S. Pacific Command

PAVN/NVA – People's Army of Vietnam/North Vietnamese Army

PIRAZ – Positive Identification Radar Advisory Zone

PRG – Provincial Revolutionary Government (shadow government of South Vietnamese communist insurgency)

PTT – Post Target Turn

RAC – Regional Assistance Command (American military advisory command for Corps Tactical Zones)

RIO – Radar Intercept Officer (U.S. Navy)

RVN – Republic of Vietnam (South Vietnam)

RVNAF – Republic of Vietnam Air Force (South Vietnam)

SAC –Strategic Air Command (U.S. Air Force)

SAM – Surface to air missile

SW – Strategic Wing (U.S. Air Force)

TAW – Tactical Airlift Wing

TPJMC – Two-Party Joint Military Commission

VA – Attack Squadron (U.S. Navy)

VF – Fighter Squadron (U.S. Navy)

VPAF – Vietnam People's Air Force (North Vietnam)

VWP – Vietnamese Workers' Party (North Vietnam)

WSO – Weapons Systems Officer (U.S. Air Force)

Contents

1

STATE OF PLAY

January 1972

As 1972 dawned, the United States' long bitter struggle to prevent North Vietnam from conquering the South was at last nearing its end. The war had so far cost more than 57,000 U.S. killed and another 300,000 wounded over nearly 20 years of American involvement. Vietnamese casualties on both sides were many times that number. Meanwhile, thousands of prisoners of war were held by both sides, including hundreds of American POWs languishing at the hands of their communist captors. Many hundreds more were listed as missing in action. President Richard Nixon had entered office in 1969 pledging an "honorable peace" in Vietnam, one that would allow the United States to withdraw while honoring its pledges to South Vietnam.[1]

To that end, Nixon's policy of Vietnamization was proceeding apace. The plan aimed at transferring responsibility for the war over to South Vietnam by simultaneously withdrawing American forces while training and equipping Saigon's military to carry on the fight. U.S. troop strength had plummeted from its peak of nearly 550,000 personnel in 1969 to just over 150,000 by the start of 1972, its lowest level since early 1965. Only about a third of those could be classified as "combat troops," however, and the administration had announced earlier in 1971 that U.S. forces would no longer engage in offensive ground operations. Further, Nixon announced on 13 January that another 70,000 U.S. personnel would be gone by 1 May. Meanwhile, South Vietnam's military had swelled to more than 1 million men under arms, although about half were regional militia and local forces. Saigon also had about 1,000 aircraft and another thousand or so naval vessels for the fight against North Vietnam and its southern proxies, the National Liberation Front (NLF). For years the NLF had been known colloquially as simply the "Vietnamese communists" or Viet Cong (VC).[2]

On the diplomatic front, Nixon announced in a nationally televised address on 25 January that, beyond the public peace talks begun with North Vietnam in May 1968, his administration had been engaged in secret negotiations with Hanoi since August 1969. This supposedly allowed both sides to discuss and "take positions free from the pressure of public debate." But both public and private talks had yielded very little progress. Nixon said he hoped to break the deadlock by going public about the secret negotiations. The president intimated that the North Vietnamese had not negotiated in good faith, on one hand publicly accusing his administration of recalcitrance and bad faith, while on the other receiving his many peace overtures in private. National Security Advisor Henry Kissinger and North Vietnamese negotiators had by that January held a dozen covert sessions in Paris. Perhaps the most significant Nixon overture came in his 11 October 1971 private message to Hanoi in which he presented his "Eight Points" plan and asked for another secret meeting to discuss. The proposal called for a ceasefire throughout Indochina, the withdrawal of all U.S. and allied forces from South Vietnam within six months of a signed agreement, and the exchange of all military and civilian prisoners of war. Other notable provisions—and downright vexing from Hanoi's perspective—called for an end to outside military infiltration into any Indochinese nation, with regional disputes to be settled based on mutual respect for territorial sovereignty and without foreign interference. Finally, Nixon's plan called for elections within six months to decide the Saigon government's fate. The elections would be open to all political groups in South Vietnam, including members of the Provisional Revolutionary Government (PRG), the shadow organization purportedly governing communist-controlled areas of South Vietnam. The PRG was created by the NLF in June 1969 to bolster the insurgency's political legitimacy both within South Vietnam and the international arena. It was comprised of high- ranking members of both the NLF and the People's Revolutionary Party (PRP), the southern iteration of the North's ruling Vietnamese Workers' Party (VWP).[3] Nixon's proposal insisted that political issues be decided only by the people of South Vietnam, free from outside meddling. Any reunification between North and South was to be accomplished through negotiation and mutual agreement between their respective governments.[4]

Hanoi at first agreed to Nixon's request for a secret meeting in November but called it off, citing the illness of Le Duc Tho, one of North Vietnam's most powerful political figures. It was Tho with whom Kissinger had already secretly conferred on multiple occasions since their initial meeting on 21 February 1970. But Hanoi neither rescheduled the meeting nor responded to Nixon's proposal. Communist leadership had for years steadfastly refused to entertain any ceasefire agreement that did not include, among other demands, the unconditional withdrawal of all U.S. forces from South Vietnam and the immediate resignation of the "puppet" South Vietnamese President Nguyen Van Thieu, who had held the office since his election in September 1967. Although incensed at Nixon's disclosure of the secret negotiations—including his attempt, in their view, to blame Hanoi for the peace talks' impasse—communist leadership tried to turn the tables on the U.S. president. With Hanoi's support, Nguyen Thi Binh, co-head of the PRG delegation in Paris, on 2 February announced that the southern communists were now willing to link the release of POWs to the withdrawal of U.S. military forces from South Vietnam. While still demanding Thieu's immediate resignation, the PRG relented on its demand that his government also be dissolved. Hanoi publicly praised Binh and the PRG for their "conciliatory gesture and eagerness to help the peace process along." Still, Hanoi rejected Nixon's televised peace offer. Although the president's offer of a concrete timeline for U.S. withdrawal was something communist leaders had long insisted upon, they remained unmoved. For the time being, there would be no further covert peace talks.[5]

While Hanoi had already offered during the secret sessions to link POW release to a concrete schedule of American troop withdrawal, the PRG concession was nevertheless of keen interest to the Nixon administration, which had made the issue a cornerstone of its negotiating stance. Prior to 1969, President Lyndon Johnson's state department, the primary agency and so-called "single spokesman" on the POW/MIA issue, pursued what it termed "quiet diplomacy" to achieve administration ends. Public statements and official press releases generally kept mention of American POWs and MIAs to a minimum, and the government encouraged servicemember families to follow suit. The stated rationale for keeping quiet held that widespread publicity might complicate matters by provoking political agitation at home while endangering the safety of servicemembers held in Southeast Asia. Further, the communists' reluctance to negotiate on the issue—indeed, in some cases to even admit that they held American POWs in the first place—required officials to work indirectly via diplomatic back channels like the International Committee of the Red Cross.[6]

That all changed when the Nixon administration initiated its Go Public Campaign in April 1969. Incoming Secretary of Defense Melvin

Laird, who had for years followed the issue and become a staunch POW advocate, believed the previous policy to be a mistake. Rather than working quietly behind the scenes, the U.S. should employ all means of public and political pressure in a full-throated demand that North Vietnam and its Southeast Asia clients abide by the 1949 Geneva Convention Relative to the Treatment of Prisoners of War. This, Laird said, must include the "basic rights" of prisoners to decent and humane treatment and communication with their families, a forthright accounting of all POWs held, and impartial inspections of communist detention camps. Not surprisingly, Laird found stalwart allies among the military services, which called for "a more vigorous public affairs and propaganda effort" on the POWs' behalf. Advocates believed that such a program, bringing to bear the full resources of the U.S. government, could "influence world opinion to the point that Hanoi will feel compelled to afford proper treatment to U.S. PWs."[7]

The status of some 1,300 U.S. servicemembers listed either captured or missing in action was almost completely unknown because of North Vietnam's refusal to provide information. Although the Democratic Republic of Vietnam had acceded to the Conventions in 1957, the communist high command maintained that, because the United States had not officially declared war in Southeast Asia, captured servicemembers were merely "criminals" or "air pirates" and thus not entitled to any of the protections afforded by the Geneva agreements.[8]

That position appeared to change somewhat when in August 1969 Hanoi released three POWs in an apparent public relations ploy. At the time, POWs, as a matter of code and honor, had agreed that none would accept early release unless all other prisoners were set free, as well. But after five years of abuse, hunger, and degradation, senior POW leadership decided the time had come to let America—and the world—know what was happening to them. Those released were USAF Capt. Wesley L. Rumble, Navy Lt. Robert F. Frishman, and Seaman Douglas B. Hegdahl. During an intelligence debriefing, Hegdahl, who prison authorities had mocked as "the incredibly stupid one"—and thus unlikely to cause trouble for Hanoi—was able to convey the names, information, and status of 256 POWs. The 22-year-old seaman had used the nursery rhyme "Old MacDonald Had a Farm" as a memory aid to help remember and deliver remarkably accurate information. On 2 September, Frishman and Hegdahl appeared at a joint press conference to testify about POW conditions. Frishman reported that his North Vietnamese captors made clear they expected him to parrot their claims that U.S. POWs were being well treated, even issuing a veiled threat that he should "not forget that they still have hundreds of my buddies in their hands." But Frishman testified that his fellow POWs had urged him to tell the truth without regard for communist reprisals. Frishman and Hegdahl went on to relate a horror story of torture, months-long solitary confinement, starvation, coerced statements and confessions, and woefully inadequate medical care for the sick and injured. The men reported that what care was available seemed designed to merely keep the POW alive rather than bring about his recovery. Following these revelations, the Nixon administration, with assistance from such groups as the National League of Families of American Prisoners and Missing in Southeast Asia, significantly ramped up its public pressure campaign. As noted, the administration placed the timely release of America's POWs at the center of its negotiating stance. After all, there could be no "honorable peace," as Nixon had often pledged, unless America brought her POWs home.[9]

But neither Nixon's disclosure of the secret talks nor his focus on the POW issue were Hanoi's main concerns in January 1972. Communist leadership was much more worried over the U.S. president's overtures to North Vietnam's most important patrons—

China and the Soviet Union. In an attempt to exploit the ever- widening Sino-Soviet split—most recently punctuated by a monthslong series of bloody border clashes between the communist giants in March 1969—Nixon set about a policy of rapprochement with China and détente with the Soviet Union. Aside from a welcome lessening of Cold War tensions, Nixon also believed that, as each communist nation sought better relations with the United States to strategically balance the other, he could leverage that desire to encourage Peking and Moscow to pressure Hanoi into negotiating peace terms acceptable to the United States. On 15 July 1971, Nixon announced in a nationally televised press conference that he would in February of the following year become the first U.S. president to visit the People's Republic of China. Likewise, Nixon would travel to the Soviet Union for a summit with its leader, Leonid Brezhnev, on 22 May 1972, becoming the first U.S. president to visit the capital city of Moscow.[10]

Hanoi feared, with some justification, that superpower concerns would trump its longtime struggle to unify the two Vietnams under communist rule. Despite its stubborn independence, North Vietnam's leadership was desperately reliant upon the billions of dollars in military, technological, and economic aid provided by its patrons. At the 27 January to 2 February Twentieth Plenum of the Vietnam Workers' Party in Hanoi, leaders painted a grim picture of the economy: although agricultural collectivization and rural electrification had improved outlook, the U.S. intensified bombing campaign up to the 20th parallel (in response to Hanoi's escalated attacks in Laos, Cambodia, and South Vietnam's northern panhandle) rendered substantial progress unlikely for the foreseeable future. Indeed, concluded Party leaders, it was becoming impossible under current conditions to build socialism in the North while fighting in the South. Doubtless, Hanoi's worries over Nixon's political overtures to China and the Soviet Union only exacerbated its concerns.[11]

On the other hand, Party leaders were quite bullish on the country's battlefield prospects. Facing a possible betrayal by its great power allies and with the Americans on their way out, Hanoi began to see an all-out, win-the-war now push as its best chance for victory. South Vietnam's botched Operation Lam Son 719 the previous year only added fuel to the fire. An outnumbered and outgunned ARVN force—though backed by American air power and logistical support—had been badly mauled in its effort to interdict the Ho Chi Minh Trail in southeastern Laos. North Vietnam's powerful General Secretary Le Duan, who had succeeded Ho Chi Minh (born Nguyen T'at-Than) as Party leader in 1960, along with deputy and principal ally Le Duc Tho, believed that North Vietnam's military had broken the back of ARVN in Laos. Vietnamization, it seemed, had been a failure. The "Brothers Le" now detected a definitive military advantage for the People's Army of Vietnam (PAVN). The time had come for decisive action.[12]

And there was little to stop them. Though Ho Chi Minh died in September 1969, he had become little more than a spiritual figurehead in his final years. Since ascending to general secretary, Le Duan and his allies had gradually accrued power and worked to marginalize those who shared Ho's belief in a more patient strategy for bringing the South under communist rule. Chief among the naysayers was the revered Gen. Vo Nguyen Giap, mastermind of the Viet Minh's signature 1954 victory over the French at Dien Bien Phu. Like Ho, Giap had long favored a protracted guerilla war to eventually drive the Americans out. The general had vehemently opposed the 1968 Tet Offensive as unnecessarily reckless, and in the intervening years had become ever-more politically isolated in Hanoi. By 1972, Le Duan and his allies had so fully solidified their grip on power that detractors even of Giap's stature could do little more than look on.[13]

What followed was the return of Le Duan's General Offensive-

General Uprising (GO-GU) strategy, which Hanoi had intermittently employed since 1965. GO-GU was a two-pronged effort. First, the general offensive phase would see the PAVN engage and destroy large enemy formations in conventional set-piece battles, gobbling up territory and bringing maximum pressure to bear across a large battlefront. Such dramatic victories, coupled with political efforts behind enemy lines, were to designed to inspire South Vietnam's population to rise up and help overthrow the Saigon government. With its "puppet" regime vanquished, the United States would have little choice but to immediately and unconditionally withdraw from South Vietnam. Le Duan and Hanoi had dramatically employed GO-GU during the 1968 Tet Offensive and its follow-on waves, a series of campaigns that most observers now recognize as a stunning propaganda victory for the communists on one hand and a stinging military defeat on the other. Of the estimated 124,000 combat troops and guerillas committed by North Vietnam and its NLF proxies in the South, some 40,000 had been killed in action, along with another 10,000 civilian conscripts, sympathizers, and political cadres. Since then, North Vietnam had scaled back its conventional offensive operations, opting instead for an "economy of forces" strategy that saw a return to guerilla tactics and smaller-unit actions. Still, despite the earlier losses—and absent the hoped-for general uprising—Le Duan and his acolytes were now ready to give it another go.[14]

The decision was only reinforced following Nixon's China visit from 21 to 28 February. Chinese Premier Zhou Enlai traveled to Hanoi on 3 March to reaffirm China's commitment to North Vietnam while also warning the VWP politburo to abandon GO-GU and instead negotiate a diplomatic end to the war. If North Vietnam pursued a military victory, Zhou said, the American president would surely "punish" it. Le Duan was incensed, accusing the PRC of "rescuing" Nixon from the failures of Vietnamization and Lam Son 719 abroad and his political vulnerability at home. Indeed, party leadership planned to use the electoral calendar against Nixon in an election year, while also leveraging congressional and domestic anti- war sentiment to blunt the president's response. When push came to shove, the politburo believed, the Soviets and PRC would rally to their socialist ally, offering a strong public rebuke if Nixon's reprisals dared strike above the 20th parallel. Ignoring Zhou, Le Duan gave the final go-ahead for the 1972 Nguyen Hue Offensive. As North Vietnam's Central Military Commission assembled the military, logistical, and political assets for what would become known in the West as the "Easter Offensive," peace in Vietnam would have to wait. The dogs of all-out war would soon slip their bonds once more.[15]

2

SHARPENING THE SPEARS

1 February – 29 March

The North Vietnamese had begun prioritizing a military rather than diplomatic solution well before Nixon's February 1972 China trip. Analyzing the results of Saigon's Operation Lam Son 719 in southeastern Laos and the relative "success" of the PAVN counteroffensive—dubbed the Route 9- Southern Laos campaign—the VWP politburo in May 1971 announced that the time had come to "develop our strategic offensive posture in South Vietnam to defeat the American 'Vietnamization' policy, gain a decisive victory in 1972, and force the U.S. imperialists to negotiate an end to the war from a position of defeat." That the U.S. had already withdrawn the bulk of its forces—with further announced cuts in the offing—along with Nixon's political vulnerability in a presidential election year, only reinforced Hanoi's resolve.[1]

Certainly, Lam Son 719 had not gone as well as Washington and Saigon had hoped. Seen by the Nixon administration as an important test of Vietnamization, the president instructed military planners to work through every detail "in order to make sure the Laotian operation works out." On 7 February 1971, Saigon ordered an infantry and armored task force, some 16,000 ARVN, Ranger, Airborne, and Marine troops backed by American air power and logistical support, across the Laotian border and down Route 9 toward the town of Tchepone about 25 miles to the west. Allied intelligence suspected the town and its environs of being a major hub of PAVN activity within Base Area 604 of the Ho Chi Minh Trail. The plan was to capture the town, disrupt and destroy PAVN infiltration and military buildup, and hold the position until the start of the monsoon season in May. Unfortunately from the allied perspective, communist spies in Saigon, along with numerous press reports beforehand, provided Hanoi with detailed information on the battle plan, from South Vietnamese troop strength to its exact invasion route and objectives. Desiring to both defeat the incursion and to humiliate and discredit Saigon and Vietnamization, Hanoi poured in powerful, well-armed and supplied infantry, armor, and artillery units to lie in wait. By early March, the North Vietnamese force had swelled to some 36,000 strong, outnumbering and outgunning the invasion element. Faced with a well-prepared communist response and severe command and control problems of its own, the invasion force quickly bogged down. Despite fierce U.S. close air support, the communists inflicted heavy casualties as the task force slogged its way west. The South Vietnamese finally captured the largely abandoned Tchepone on 6 March using two airlifted battalions of the ARVN 1st Division. But Hanoi soon launched a powerful counteroffensive to chop up and destroy ARVN formations and artillery firebases in the area. With the situation growing more dire by the day, South Vietnamese President Thieu had seen enough. On 9 March, he ordered Tchepone abandoned and all South Vietnamese forces withdrawn east through a murderous gauntlet of prepared ambushes and artillery bombardment. The operation officially concluded on 6 April.[2]

Despite heavy losses on both sides, each claimed victory. Saigon and Washington said the operation had killed some 15,000 PAVN troops, while knocking out large numbers of tanks, artillery, and other heavy weapons. The allies further contended that the incursion had disrupted an imminent communist offensive by seizing tons of war materiel and denying the area as a staging ground. But the campaign had come at a heavy price. U.S. estimates reported some 9,000 South Vietnamese killed, wounded, or missing in action. Although the U.S. did not participate in the ground assault, it lost 215 killed and hundreds wounded through its support and advisory role. U.S. and South Vietnamese air losses were staggering, with more than 100 helicopters destroyed and another 600 damaged. Further losses included several fixed-wing close air support fighter- bombers, about 200 tanks, armored fighting vehicles, and half-tracks, and some 115 artillery pieces destroyed or captured. Most importantly from the perspective of Vietnamization, the operation demonstrated that South Vietnamese forces, still heavily reliant upon American airpower, logistics, and advisers, were not yet ready for primetime. Despite its own heavy losses, Hanoi looked upon the results and saw opportunity. With the Americans on the way out, and Vietnamization faltering, the time seemed right for another all-out push to win the war.[3]

In June, the Central Committee of the VWP politburo approved the offensive campaign dubbed "Nguyen Hue," which honored the birth name of one of Vietnam's most storied military commanders, Emperor Quang Trung, who defeated a Chinese invasion in the late 18th century. The plan called for a three-pronged offensive that would see powerful conventional thrusts along multiple axes of advance: across the DMZ and Laotian border into South Vietnam's northernmost Quang Tri Province, another into the Central Highlands province of Kontum from communist sanctuaries in Laos and Cambodia, and a third from Cambodian base areas west of Tay Ninh and Binh Long provinces northwest of Saigon. Each theater commander would have at his disposal three to four infantry divisions, along with armor, artillery, and other supporting arms. In keeping with Le Duan's general offensive-general uprising doctrine, the invasion's near-term objectives were two-fold. First, PAVN conventional forces were to annihilate South Vietnamese regimental and brigade task forces, ideally rendering entire enemy divisions combat ineffective. PAVN forces were to seize and hold new ground as they went, solidifying existing base areas while grabbing vast swaths of South Vietnamese territory. At the same time, North Vietnam, in coordination with its southern communist allies, was "to provide direct support for mass popular movements that would conduct attacks and uprisings to destroy the pacification program in the rural lowlands" and further disrupt and destabilize the Thieu government.[4] The success of the U.S.-South Vietnamese pacification effort known as Civilian Operations and Revolutionary Development Support (CORDS) had proved frustrating and alarming for communists North and South. Begun in May 1967, by March 1972 the program led MACV's Hamlet Evaluation Survey to conclude that some 97 percent of South Vietnam's rural population was living in "totally or relatively secure" villages—a dramatic improvement over the previous decade and one that did not bode well for the communist insurgency.[5] Finally, in anticipation of at least limited American reprisals above the 20th parallel, the plan called for strengthening Northern military and self-defense capabilities to "be prepared to defeat any reckless operation the enemy might launch." A major step was to reverse the demobilization of the National Defense forces begun following the Johnson administration's November 1968 bombing halt.[6]

By July, the politburo's collective mind was made up. "The time has come for us to bring about a favorable moment. We must intensify our struggle, energize our military…attacks. We must generalize our assaults in the South and in the rest of Indochina. It is essential that we crush America's efforts at 'Vietnamization'…to achieve a decisive victory in the year 1972 and compel the American imperialists to end the war through negotiations on our terms." This was entirely consistent with Hanoi's well-worn "talk-fight" strategy.

VWP leaders had long attempted to wring political concessions at the bargaining table via military victories on the battlefield. A triumphant spring offensive would further discredit Vietnamization and the Thieu regime, bring to a boil the always-simmering American anti-war protest movement, and generally weaken Nixon's negotiating position as he faced a hostile Congress and his own reelection bid in 1972. Thus, VWP leadership instructed its delegates at the Paris peace talks in July 1971 to offer no further compromises with the Americans. The course was set.[7]

For the next six months, commanders put PAVN forces through their training paces, prioritizing the multi-regimental and divisional coordination necessary for a massive conventional invasion. Anticipating South Vietnamese tactics and disposition, special attention was paid to overcoming static, fortified strongholds, and defensive positions. In November 1971, the 308th PAVN Infantry Division was charged with constructing a mockup of an enemy regimental base camp, complete with an interconnected system of blockhouses, bunkers, trenchworks, obstacles and so on. Once completed, the 308th was ordered to overrun and annihilate the "enemy" camp in a four-day training exercise. The lessons learned were recorded and transmitted to other units. The division was also tasked with the live-fire testing of newly acquired Soviet weapons, including the wire-guided *Malyutka* anti-tank missile system. Dubbed the AT-3 Sagger by allied forces, the remote-controlled system was highly sophisticated for the time. Its operator used a long-range "periscope" and joystick to guide the six-pound warhead on target. Capable of penetrating eight-inch armor, enough Saggers were provided by the Soviets to outfit three PAVN anti-tank companies, all of which were sent to the DMZ theater.[8] Another potential game-changer was the newly arrived *Strela*-2 Man Portable Air Defense system, a shoulder-fired anti-aircraft missile codenamed the SA-7 Grail by NATO. During Egypt's so called "War of Attrition" with Israel in 1969, the Grail was the first man portable anti-aircraft missile successfully used in combat. It would soon gain a fearsome reputation among allied pilots in Vietnam. Fortunately from the anti-communist perspective, the SA-7's propensity of locking on to any heat signature blunted its effectiveness once pilots discovered the tendency and adjusted tactics.[9]

With military training underway, the politburo next looked to shoring up the invasion force's political morale. Despite the formidable array of infantry, armor, artillery, anti-aircraft, and new Soviet weaponry massing along South Vietnam's borders, Hanoi's leaders knew they faced a daunting challenge—especially at the hands of still-potent U.S. airpower. The Central Military Party Committee therefore instructed all units up to division-level to hold Party congresses to explain the politburo's rationale "to raise combat morale and increase our troops' resolve to overcome adversity and endure the challenges of savage combat…to encourage them to fight on until complete victory was secured."[10]

By late 1971, logistical preparation of the battlefield was fully underway. Transportation Group 559, the unit that since 1959 had been responsible for creating and maintaining the Ho Chi Minh Trail in eastern Laos and Cambodia, had worked steadily to expand the Trail's road and oil pipeline system. By the start of the dry transportation season in late 1971, the official communist history claimed that an all-time high of some 8,000 transportation trucks were operating along the North's "strategic transportation corridor," up 2,000 over the previous year. By early 1972, North Vietnam had doubled the supplies sent south over the previous year, including more than 10,000 tons of oil and gasoline to fuel the coming mechanized push into South Vietnam—all of this despite a concerted and ongoing allied aerial interdiction campaign all along the Trail and other staging areas. A secondary and tertiary road system was constructed by engineer troops, civilian laborers, and others to ferry supplies from the Trail to unit assembly areas near their lines of departure. Additionally, "warehouses" were built to further stockpile weapons, food, and medical supplies close to invasion routes. Again according to the Party's official history, a massive "rice campaign" was undertaken to purchase, seize, and stockpile thousands of tons of rice to feed invasion troops. Conscripted civilian laborers then hauled the supplies from the lowlands to communist staging areas in the mountains west of South Vietnam.[11]

Still, when the politburo met in March 1972 for a final pre-launch assessment, VWP leadership noted several areas of concern. These included "the limited ability of our main force units to conduct combined-arms combat operations and to attack solid, fortified defensive positions," a vitally important capability for successful corps-level operations against a defensive-oriented enemy. The politburo also complained that some theaters "had been slow at completing supply preparations and road construction" and that "the quantity and quality of our local political and armed forces had not grown rapidly enough to enable them to carry out their duties in this campaign."[12] Such admissions in an official Party document were extraordinary and may represent a retroactive attempt to explain away some of the offensive's ultimate failures. It may also explain why VWP leaders ordered an eleventh-hour change in plans. Initially, the DMZ-Quang Tri sector, dubbed the "Tri-Thien" theater by Hanoi, was merely to be a "supporting" front, with the main efforts coming in the Central Highlands and the "eastern Cochin China" region northwest of Saigon. But following the March 1972 review, the CMPC reversed course, designating Tri-Thien as the primary invasion route, with the eastern Cochin China and the Central Highlands fronts relegated to division-strength "deep penetrations" only if the opportunity presented itself. Doubtless, committee planners' disappointment with preparations in the latter two theaters, along with Tri-Thien's proximity to the better-developed logistics networks, SA-2 surface-to-air missile (SAM) sites, and even MiG fighter bases in North Vietnam's southern panhandle made it the preferred choice to lead the way.[13]

The CMPC created a Party Committee and Campaign Command for the Tri-Thien front and appointed Maj. Gen. Le Trong Tan commander. By mid-March, units reached their assembly areas in all three theaters. Tan had at his disposal three infantry divisions, three standalone infantry regiments, two "composite" anti-aircraft divisions featuring eight anti-aircraft artillery and two missile regiments, nine field artillery regiments, two tank and armored regiments, two engineer regiments, and 16 battalions of sapper, signal, and transportation troops. In the Central Highlands, the North Vietnamese amassed two infantry divisions, four standalone infantry regiments, five artillery, engineer, and sapper regiments, six anti- aircraft battalions, and one tank battalion. Finally, the eastern Cochin China front northwest of Saigon saw another three infantry divisions, four standalone infantry regiments, along with another four regiments and eight standalone battalions of specialty branch troops, including two field artillery regiments and a tank battalion. All told, North Vietnam's invasion force stood at some 225,000 troops, with hundreds of Soviet-made PT-76 and T-54/55 tanks, 122-mm rocket and 130-mm heavy artillery, along with a formidable mix of self- propelled ZSU-57-2 autocannons, SA-7s, and other anti-aircraft assets. Known in the West as the "Easter Offensive," Nguyen Hue was by far the largest, most powerful military array ever assembled by North Vietnam. But all that combat power would come at a cost, severely testing its commanders' ability to conduct corps-level, combined-arms operations across three strategic battlefronts. How would they meet the challenge? And how would the South Vietnamese and Americans respond to the onslaught? The answers were not long in coming.[14]

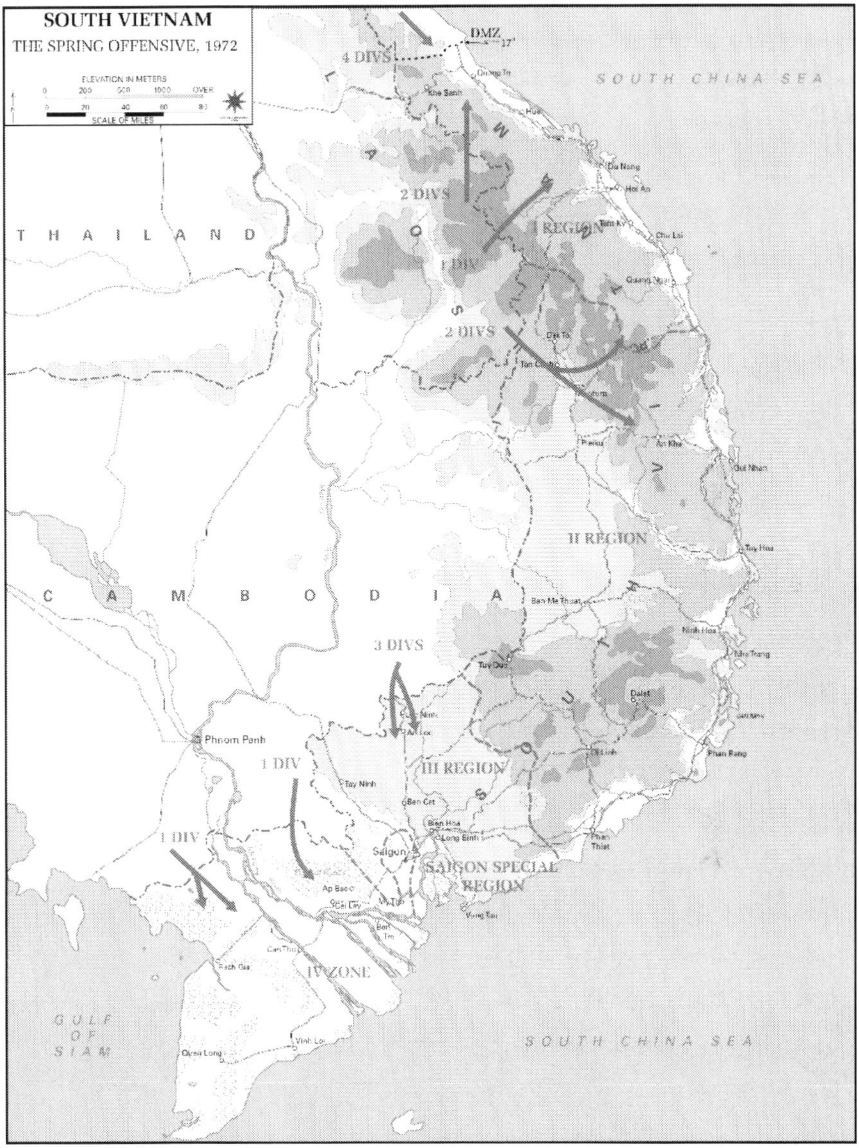

Figure 1: The 1972 Nguyen Hue Offensive

3

ACROSS THE RUBICON

30 March – 9 May

As dawn broke on 30 March, South Vietnamese I Corps commander Lt. Gen. Hoang Xuan Lam had some 25,000 troops at his disposal in South Vietnam's northernmost provinces of Quang Tri and Thua Thien. Lam, who had had primary responsibility for coordinating Operation Lam Son 719 a year earlier, counted among his force elements of two ARVN infantry divisions, two Marine brigades, and an assortment of local territorial troops scattered across the area. Many of Lam's troops occupied a series of outposts and artillery firebases spread out in a jagged line stretching west from the coastal plain to a cluster of firebases and strongpoints near the Laotian border. By midday, South Vietnamese positions came under heavy attack by communist long-range artillery north of the DMZ, effectively suppressing ARVN counter-artillery fire and support. Soon after, some 30,000 PAVN troops, protected by a blanket of low cloud cover that hindered allied airpower, flowed across the DMZ and from eastern Laos, some launching from staging areas abandoned following the Lam Son 719 withdrawal. Over the ensuing 48 hours, government positions toppled like dominoes, as the poorly integrated and supported strong points were quickly cut off and overrun. Perhaps the most devastating initial loss came on 2 April with the sudden surrender of Camp Carroll, the defensive line's western anchor and home to the bulk of I Corps heavy artillery. The capitulation handed over to the advancing 308th

NVA Division some 1,500 men of ARVN's 56th Regiment, 3rd Division, including several mammoth 175 mm guns, imperiling the South Vietnamese defensive position west of the provincial capital of Quang Tri City. Only the timely arrival of some 9,000 Marine and Ranger troops released from the strategic reserve allowed Brig. Gen. Vu Van Giai, commander of what was left of the ARVN 3rd Division, to establish a new defensive line west of Dong Ha and Quang Tri City on 8 April. South Vietnamese defenses in western Quang Tri, however precarious, had for the moment stabilized.[1]

Meanwhile east along the coastal plain, the 3rd Division's 57th Regiment was falling back in the face of a division sized NVA force pushing hard along Highway 1 to cross the junction of the Cua Viet and Mieu Giang rivers at the Dong Ha Bridge just north of town. If the NVA could cross in strength, the road to Quang Tri—and even Hue farther south—was wide open. After a spirited defense that employed two score of newly arrived M-48A3 Patton tanks, it became clear to American advisers Capt. John Ripley and Maj. John Smock on 2 April that the weight of NVA numbers was simply too heavy to hold. The Dong Ha Bridge had to be blown. Ripley, a 32- year-old Marine would be awarded the Navy Cross—second only to the Medal of Honor for valor—for knocking out the bridge, an effort that saw him make repeated forays to rig the bridge's underside with explosives, swinging arm over arm above the Cua Viet River as NVA infantry and T-54/55 gunners desperately sought to take him down. The loss of the bridge forced the NVA to divert west for a crossing at Cam Lo, buying valuable time for Quang Tri's defenders. Still, it was only delaying the inevitable. Over the next three weeks, Hanoi continued to pour men and materiel into the Tri-Thien front. This included the first-ever placement of SA-2 anti-aircraft missile sites onto South Vietnamese territory. By 27 April, the renewed and resupplied NVA were on the move once more. Whatever fight was left in the remnants of the 3rd ARVN division evaporated, as entire units began retreating under their own volition before finally abandoning their equipment, stripping off their uniforms, and joining the civilian throng fleeing toward relative safety south of the My Chanh River. Despite I Corps commander Lam's order to "hold at all costs," Quang Tri City capitulated on 2 May, marking the first provincial capital to fall since the 1968 Tet Offensive. A few days later, the communists had secured their grip on the rest of the province.[2]

Horrifically, the glut of civilians fleeing south toward the city along Highway 1 soon fell under indiscriminate NVA shelling and ambush. The Red Cross later estimated that some 2,000 were killed, many of them women, children, and the elderly. The scene quickly came to be known as "Terror Boulevard" in South Vietnamese news reports. In September, North Vietnamese defector private first class Le Xuan Thuy, 22, would claim at a Saigon press conference that attacks on civilians were intentional. Thuy, a radio operator with the 324th Division, had been part of an ambush position south of the city. The defector said that he and his fellow soldiers were ordered to fire on anyone traveling down the road, whether civilian or military, because "those who moved southward were our enemy...and had to be destroyed."[3]

While fighting raged in Quang Tri, communist staging areas were abuzz in western Thua Thien Province's A Shau Valley to the south. The NVA 324B Division stood poised to drive east toward the ancient imperial capital of Hue City, site of so much carnage and hard fighting during the infamous 1968 Tet Offensive. By 28 April, the division's 29th and 803rd NVA regiments had overrun fire support base (FSB) Bastogne which guarded Hue's southwestern flank. The loss of Bastogne rendered secondary outposts in western Thua Thien untenable, as elements of the 1st ARVN Division were forced to withdraw east. By 2 May, the 324B Division brought ARVN defensive positions on the high ground west of the Perfume River under murderous rocket and artillery fire. Meanwhile, the 304th NVA Division pushed against the still forming My Chanh River defensive line from the northwest. ARVN defenses in the newly formed "Hue Pocket" were now in danger of being crushed on three sides by thousands of NVA troops, armor, and heavy artillery.[4]

The opening rounds of Nguyen Hue had cost I Corps' northern command dearly. The 3rd ARVN Division had simply ceased to exist, shattered beyond the point of merely replacing men and materiel to restore its combat effectiveness. Meanwhile, South Vietnamese Marines had taken approximately 1,000 casualties, while another 1,500 Rangers were listed as killed, wounded, or captured. The 1st Armored Brigade had suffered 1,200 casualties, accounting for nearly 60 percent of its personnel. It had also lost the bulk of its armor— desperately needed to blunt Hanoi's armored columns rolling south. Some 400 vehicles, including 43 M48 and 66 M41 tanks and 103 M-113 armored personnel carriers (APC) had either been destroyed or abandoned in the headlong flight south. Out of a force consisting of a tank regiment and three armored squadrons, just one M48, six M41s, and a few APCs made it across the My Chanh. Finally, nearly all of I Corps' heavy artillery—some 140 pieces—had either been destroyed or captured. In coming weeks, many of those captured American-built 105 mm and 155 mm howitzers would be turned against South Vietnamese defenders. Now, Hue City, just 35 miles south of Quang Tri, was once more in communist crosshairs. Whether the South Vietnamese would have the men, materiel, and moxie to deny the North that cherished prize remained to be seen.[5]

§

To the south, NVA forces had launched their assaults on the eastern Cochin China front, as well. III Corps commander Lt. Gen. Nguyen Van Minh had about 40,000 troops with which to defend the region's 11 provinces, including the Saigon Special Zone around the capital. Minh had positioned many of his men in a series of defensive positions and strongpoints in western Tay Ninh and Binh Long provinces. While this passive strategy allowed for defense-in-depth, it did very little to thwart the prodigious communist buildup across the Cambodian border. And it was indeed formidable. Lt. Gen. Tran Van Tra had three PAVN divisions and three independent regiments, along with armor and artillery, poised to strike. On 2 April, the NVA rolled across the frontier at Tay Ninh, gobbling up ARVN outposts along the way. Fearing that Tay Ninh City was vulnerable, Minh, in consultation with 3rd Regional Assistance Command (RAC) adviser Maj. Gen. James F. Hollingsworth, ordered outlying elements of the ARVN 25th Division to pull back and reinforce the city. But Tra had other plans. His primary objective was actually neighboring Binh Long Province to the east and its capital of An Loc. There, Hanoi hoped to establish a PRG seat of power, menacing Saigon just 50 miles south on Highway 13. A victory at An Loc would produce not only an undeniable psychological victory but would discredit the Thieu regime and Nixon's Vietnamization strategy.[6]

The district capital of Loc Ninh, about nine miles southeast of the Cambodian border, was garrisoned by the 9th Regiment, 5th ARVN Division and its small American advisory team. By 5 April, the town fell under heavy attack, as the entire 5th NVA/VC division, along with the 203rd Tank Regiment, unleashed a ferocious armored and artillery assault. Despite clearing skies and two days of relentless pummeling by Republic of Vietnam Air Force (RVNAF) close air support, American-piloted A-37 Dragonfly fighter-bombers, and AC- 130 Spectre gunships from the 8th Special Operations Squadron out of nearby Bien

Hoa Air Base, the communists would not be denied. On 7 April, the town fell, with the NVA taking some 1,300 prisoners—including South Vietnamese regimental commander Col. Nguyen Cong Vinh—and capturing two batteries of 105 mm and 155 mm howitzers. A few survivors managed to escape south to the relative safety of An Loc about 12 miles down Route 13, but senior ranking American adviser Lt. Col. Richard S. Shott perished as NVA armor overran his position. Master Sgt. Howard B. Lull, initially believed to have escaped and evaded capture, would later be officially listed KIA, as well. At least three other American advisers would be captured and held in Cambodia. The northern approaches to An Loc now lay open.[7]

Responsibility for its defense fell to Brig. Gen. Le Van Hung, commander of the ARVN 5th Division. Hung had at his disposal some 7,500 troops of varying quality, 2,000 of which were territorial militia, along with a few 105 mm and 155 mm artillery pieces. He was assisted by U.S. Army Col. William Miller and another two dozen American advisers. Hung soon found himself outnumbered more than four to one, as three NVA divisions, an additional artillery division, two armored regiments, and other supporting units coiled into the vicinity of An Loc to deliver the final blow. Elements of the 7th NVA Division moved to set up strong blocking positions at Tan Khai and Tan O south on Highway 13. Over the coming weeks, the 1st Airborne Brigade would first try to fight its way up the highway to clear the positions and was later followed by the 21st ARVN Division's own attempt. Neither were successful. By mid-April, the city had been completely cut off. It was now wholly dependent upon the U.S. 374th Tactical Airlift Wing (TAW) out of Bien Hoa for resupply and reinforcement. Enemy commanders anticipated the move and beefed up anti-aircraft assets around the city. By the end of May, three big C-130 transports and another AC-119K Stinger gunship had been shot down. Over the next two weeks the NVA tightened the vice, as artillery and mortar fire—

including recently captured South Vietnamese guns—saturated the city. Several combined infantry and armored thrusts managed to gain footholds in the city's northeastern and western outskirts. Only a resolute defensive stand, stiffened by M72 anti-tank rocket fire and merciless close air support directed by Col. Miller, kept the NVA from advancing further. Enemy troops fell by the hundreds, along with about two dozen tanks destroyed. Farther out, B-52 Arc Light strikes pummeled suspected enemy formations to within a half mile of the city's defensive positions. On 19 April, the NVA launched its all-out assault. After a preparatory rocket, mortar, and artillery barrage of some 1,000 rounds, the NVA's 9th Division made some headway but was ultimately driven back yet again in heavy house-to-house fighting and tactical air strikes. Additionally, the timely redeployment of the ARVN 21st Infantry Division from the Mekong Delta to Chon Thanh south of An Loc had forced the NVA to guard against an attack to its rear, further thwarting the main assault. After losing an estimated 2,000 troops KIA, along with numerous mechanized assets, the NVA's assault finally ground to a halt. By 23 April, invader and defender alike were exhausted. As each side desperately worked to solidify its position, the attack on An Loc settled into siege.[8]

§

Some 300 miles northeast in the Central Highlands province of Kontum, the NVA's 320th Division under Maj. Gen. Hoang Minh Thao launched the third prong of Hanoi's Nguyen Hue campaign on 12 April. The attack on South Vietnam's II Corps had actually begun a week earlier when the NVA's 3rd Infantry Division under Maj. Gen. Chu Huy Man, augmented by sapper battalions and local force Viet Cong, struck from base areas in the An Lao River valley of Binh Dinh

Province along the coast. Taken by surprise, II Corps commander Lt. Gen. Ngo Dzu—who had expected the first blows to land in the northwest Central Highlands—scrambled to hold the line as Man's forces pushed south and east toward Hoai Nhon, Hoai An, and the all-important north-south thoroughfare, Highway 1. Now came Thao's bid to finally realize Hanoi's years-long ambition of splitting South Vietnam in half along an east-west axis through northern II Corps. The first to fall on 21 April was a series of South Vietnamese fire support bases and strongpoints along the "Rocket Ridge" high ground east of Highway 14 halfway between Ben Het in the north and the provincial capital of Kontum in the south. Two days earlier, some 20,000 NVA troops of the 2nd Division, four standalone regiments, and armor from the 203rd Tank Regiment poured in from Laotian sanctuaries, bypassing Ben Het to swarm the ARVN garrisons at Tan Canh and Dak To II. With just 2,000 troops at his disposal to defend both positions, 22nd ARVN Division commander Col. Le Duc Dat did not stand a chance. After a ferocious artillery and rocket bombardment, NVA tanks led the way, followed by a full infantry assault. While defenders were equipped with some number of M41 tanks, communist infantry made quick work of them using the newly arrived Soviet AT-3 Saggers. Once again, despite merciless close air support strikes to stem the tide, both Tan Canh and Dak To II fell to the enemy by 25 April. Aside from eliminating the 22nd Division as a cohesive fighting force, the NVA also captured nearly two dozen artillery pieces and large stores of ammunition in the area. Now Kontum, an important seat of government authority, power, and resources in the northwestern Highlands, lay just 30 miles south along Highway 14. Largely unguarded, it was ripe for the picking. However, the NVA lingered and did not immediately move against the city. North Vietnam's official history of the war blamed various logistical problems for the delay, but the reasons remain unclear. At any rate, responsibility for the city's defense was handed to

Col. Ly Tong Ba, commander of the 23rd ARVN Division, whose regiments were scattered throughout the Highlands. Whether Ba would have the acumen—and the firepower—to hold the provincial capital in the face of a dawdling, yet powerful, NVA force was an open question.[9]

By early May, Hanoi's Nguyen Hue Offensive had achieved remarkable success on all three fronts. In the Tri-Thien theater (I Corps), NVA forces had captured all of Quang Tri Province and were threatening Hue City in Thua Thien Province to the south. Near Saigon, the communists had driven deep into South Vietnamese territory, bringing the Binh Long provincial capital of An Loc under murderous siege. Finally, the Central Highlands front had seen NVA forces gobble up large swaths of Kontum Province, with powerful enemy forces moving in seize its vitally important provincial capital. Throughout, the performance of ARVN troops and leadership had been mixed. Some units and commanders exacted a terrible toll on advancing NVA forces, while others offered only token resistance before retreating headlong from the fray. MACV commander Gen. Creighton Abrams warned his superiors in Washington that the situation was dire, especially in the north, where the communists were once more poised to overrun Hue as they had four years earlier. Worse, South Vietnamese military leadership was nearing its breaking point. If ARVN units lost their will to fight, Saigon's government might very well collapse.[10] Remaining U.S. airpower in the region had helped staunch some of the bleeding but could only do so much. The chaos on the ground, along with unseasonable cloud cover, rain, and fog, had worked to blunt its effectiveness. If the Nixon administration could not find an answer soon, there was little hope of preserving either the Saigon government or an "honorable" American exit from South Vietnam.[11]

To be sure, Washington's calls for a diplomatic solution had fallen on deaf ears in Hanoi. Mere hours after the fall of Quang Tri City on 2 May, Kissinger and Le Duc Tho met in Paris for the first time in eight months. Nixon's national security advisor lambasted Tho and delegation head Xuan Thuy for their country's egregious violation of the spirit of peace negotiations. Kissinger accused North Vietnam of reverting to its well-worn "talk-fight" strategy of wringing political concessions through battlefield aggression. Kissinger demanded that Hanoi halt its offensive, withdraw all its forces to their pre-29 March positions, and reenter serious negotiations to reach a settlement. The communists were unmoved. If Washington wanted peace, Tho countered, then it must agree to "the immediate resignation of Thieu [and] the adoption of a policy by the Saigon administration of peace, independence and neutrality" among other concessions. This was essentially the same list of demands the communists had made since the Johnson administration. That Tho offered no counterproposal was telling, and Kissinger later reported that it "suggested that the recent military successes had stiffened Hanoi's resolve to win the war on the battlefield." There would be no discussion of ceasefire, Tho concluded, because "the prospects for the North Vietnamese were looking good." After three hours, a frustrated Kissinger ended the discussions, with neither side requesting a future meeting. Tho and Thuy's hardline stance was applauded in Hanoi. The VWP politburo praised the men for rejecting Kissinger's entreaties outright. There were to be no further private talks pending Nixon's Moscow summit at month's end, after which "the politburo would evaluate the military and diplomatic situations."[12]

Such North Vietnamese intransigence was exactly what Nixon had feared when he embarked on his risky Vietnamization strategy nearly three years earlier. While the steady drawdown of American forces had worked to undermine some of his most ardent domestic critics, the ever-weakening U.S. military capability in Southeast Asia seemed to have both emboldened the North Vietnamese and given the president few tools with which to respond. But one important policy instrument remained: airpower. Even with Vietnamization in full swing, the U.S. at the start of 1972 still maintained about 1,000 planes in theater. Fewer than half, however, were strike aircraft. By March 1972, just three squadrons of F-4 Phantoms and a single squadron of A-37 Dragonflies remained in South Vietnam. While the A-37 was an antiquated, 1950s-era ground-attack craft, the Phantom was one of the premier air superiority fighters of its age. The F-4 was designed as a sophisticated over-the-horizon interceptor to shoot down inbound Soviet nuclear bombers with its advanced AIM 9 Sidewinder and AIM 7 Sparrow radar and laser-guided air-to-air missiles. The Phantom's top speed of nearly 1,500 miles per hour and superior climb rate could best anything in the Communist Bloc.[13]

Thailand-based assets were in better shape, with 161 Phantoms on hand, along with 52 B-52s, 16 F-105 Wild Weasel SAM suppressors, and 28 gunships. Another 31 B-52s were stationed at Guam's Andersen Air Force Base, with about 140 strike aircraft available from the aircraft carriers *Coral Sea* and *Hancock* off the coast.[14] Still, that number would have to quickly and dramatically increase if the U.S. hoped to do anything other than stave off immediate defeat. Nixon, incensed at the North's obstinance, vowed to make that happen. In the meantime, the president ordered a round of intensified bombing above and below the DMZ and in eastern Laos as a means of slowing NVA gains and relieving pressure on Hue. The president also hit the North with additional B-52 strikes, a tactic he believed to be "exceptionally effective, the best ever in the war."[15]

To be sure, the stakes for such an escalation were very high. Nixon's advisers warned of the potentially disastrous consequences of bombing too heavily and too far north of the 17th parallel. With the

Moscow summit looming, and with it talks to limit strategic nuclear weapons, the Soviets might very well cancel in protest, damaging both the prospect of détente and Nixon's effort to exploit the rift between the USSR and China. The intensified bombings would likely flare domestic opposition, as well, painting Nixon as a war monger and endangering his bid for a second term in office. But the president had made up his mind. American credibility—in the eyes of ally and adversary alike—was at stake. Achieving "peace with honor" in Southeast Asia, he concluded, was more important than better relations with the Soviets or even his own reelection. "The Summit isn't worth a damn," he later wrote, "if the price for it is losing Vietnam."[16]

Nixon announced his decision via a nationally televised address on 8 May. In light of North Vietnam's full-scale invasion of the South, along with Hanoi's refusal to discuss a ceasefire and resume good-faith peace negotiations, his administration was prepared to "do whatever is required to safeguard American lives and American honor." To that end, the president vowed to dramatically escalate the air war above the 17th parallel. Further, and in accordance with intelligence indicating that some 85 percent of Hanoi's imports—and nearly all of its Soviet-supplied oil and sophisticated military equipment—arrived by sea, Nixon announced that for the first time the U.S. would mine Haiphong harbor and other North Vietnamese ports. The idea was not new. Adm. U.S. Grant Sharp, commander-in-chief, Pacific Command (CINCPAC) from 1964 until his retirement in 1968, had advocated mining the ports, later lamenting that, "Of all things we should have done but did not do, the most important was to neutralize Haiphong."[17]

President Johnson, however, doubted its usefulness and feared such a provocation might lead to Soviet and Chinese intervention. But by spring 1972, Adm. Thomas Moorer, Chairman of the Joint Chiefs of Staff, encountered no such resistance from his commander in chief. The president tasked the Navy man with drawing up the operation. These measures, Nixon said, would remain in place until Hanoi agreed to leave the South Vietnamese to decide their own fate, to return all prisoners of war, and to accept an internationally supervised ceasefire throughout Indochina. Notably absent from the president's requirements, however, was the demand that NVA troops withdraw from South Vietnamese territory. Nevertheless, once VWP leadership acquiesced, the U.S. was prepared to halt the bombing and fully withdraw from South Vietnam within four months—two months sooner than he had offered in January. Finally, Nixon addressed Soviet leadership, placing the onus for improved U.S.- Soviet relations squarely on Moscow. The USSR, Nixon warned, must encourage its North Vietnamese client to return to serious peace negotiations and not allow Hanoi's actions—and the U.S. response to them—to threaten better relations between the superpowers. "We…are on the threshold of a new relationship that can serve not only the interests of our two countries, but the cause of world peace," he said. "We are prepared to continue to build this relationship. The responsibility is yours if we fail to do so."[18]

The course set, the United States had "crossed the Rubicon," as Nixon put it. Tactically, the new air campaign was designed to crush North Vietnam's offensive by pounding enemy troop concentrations while strangling the flow of men and materiel sustaining the invasion. The strategic objectives were even more bold: destroy North Vietnam's very ability to wage war by annihilating its industrial base to force Hanoi back to the bargaining table. Looking back, the president lamented the United States' weak response to a long list of North Vietnamese provocations, a nod to both his own past failures to answer communist aggression and to the Johnson administration's ill-conceived Operation Rolling Thunder.[19]

Begun by President Lyndon B. Johnson's administration in March 1965, the campaign was heavily influenced by then-Secretary of Defense Robert McNamara. A holdover from the Kennedy Administration, McNamara had been president of Ford Motor Company prior to entering government service and believed business techniques such as statistical and cost-benefit analyses were best suited to solving military problems. McNamara's doctrine of "graduated pressure" when dealing with North Vietnam epitomized the approach. The concept called for the application of just the right balance of carrots and sticks—diplomatic and financial overtures combined with the use of restrained airstrikes—to signal to Hanoi that the costs of trying to conquer South Vietnam outweighed the benefits. In other words, hit the enemy just hard enough to cause a little pain but be ready to halt the bombing and offer generous financial aid packages to help Hanoi "come to its senses." Therefore, under the ever-watchful gaze of the National Command Authorities (NCA)—essentially LBJ and McNamara—air commanders were restricted to targets only within specific areas of North Vietnam. Others—even those of high military value—were left untouched. "They can't hit an outhouse without my permission," Johnson allegedly once boasted.[20]

The objectives of this gradually escalating bombing campaign were threefold: damage North Vietnam's warmaking capacity and materiel stocks; restrict imported material assistance flowing from Warsaw Pact and Chinese sources; and interdict the flow of men and materiel into South Vietnam via the Ho Chi Minh Trail. However, worried over political blowback at home, escalated tensions with the Soviets, and especially fearful of direct Chinese intervention as had happened in Korea, Johnson hampered the campaign's effectiveness from the start. The process saw the Joint Chiefs—even Johnson himself—frequently intervene to handpick individual targets. Once selected, the pros and cons of hitting the target were hashed and rehashed until the ultimate decision was finally handed down. But even when commanders received the go- ahead, Johnson and his advisers further delayed by dictating the day, time, force structure, even weaponry to be used. All to presumably minimize civilian casualties and the resulting political fallout. Additionally, the administration declared downtown Hanoi and Haiphong off limits, delineating a 30 nautical mile no-strike buffer around the capital city along with a 10-mile ring around North Vietnam's largest port. This effectively created safe zones for the two most important cities to the North's war effort. Finally, the administration imposed a 30-mile no-strike zone along the Chinese border. The effect was twofold. First, the bombing injunction allowed an unimpeded flow of war materiel down the northeast and northwest rail lines from the PRC into North Vietnam. Second, by declaring these and other areas off limits, the Johnson administration inadvertently allowed North Vietnam to shift its finite air defenses from no-strike zones to areas likely to be hit, thus concentrating fire and further imperiling U.S. pilots. An added complication was Johnson's frequent bombing pauses in furtherance of his various peace overtures toward Hanoi. Most significant was the president's 31 March 1968 bombing prohibition above the 20th parallel in the wake of the initial Tet Offensive two months earlier. The president would eventually extend the bombing embargo to all North Vietnam on 1 November 1968, ending the Rolling Thunder campaign.[21]

Upon his inauguration in January 1969, Nixon continued his predecessor's policy, fearing both the collapse of the Paris negotiations and re-inflaming the antiwar movement. Both men's decisions were to have dire consequences for the fate of South Vietnam. Indeed, despite the many self-imposed limitations on Rolling Thunder, it had impacted the North's war effort, nonetheless. According to a 1967 CIA report, the campaign had cost the North Vietnamese some $215 million (about $2 billion today) in repair and reconstruction—including about $80 million in lost military equipment and supplies. Perhaps more importantly, Rolling Thunder had tied up between 600,000 and 700,000 personnel involved in everything from rebuilding and transportation to civil and air defense. As to the last, CIA analysts calculated that some 83,000 military personnel had been diverted to anti-aircraft defense alone. The resulting manpower shortage had "limited North Vietnam's capability for sustained large-scale conventional military operations against South Vietnam." The ensuing three-and-a-half-year bombing halt, however, had not only allowed Hanoi the time and security to rebuild and expand its manufacturing and logistics base, but had freed up significant manpower and material resources to wage full- scale war on the South. Secure from air attack, communist planners had thus busied themselves with moving MiG bases, SAM sites, and antiaircraft artillery close to the DMZ, while massing hundreds of tanks, artillery, and more than 200,000 troops along South Vietnam's borders.[22]

Nixon declared that it was time for a change. "I think we have had too much of a tendency to talk big and act little," he told his advisers. "We have warned the enemy time and time again and then have acted in a rather mild way when…[he] tested us." But now Hanoi had crossed its own Rubicon, and the time had come for bold action. "We have the power to destroy his war-making capacity," he continued. "The only question is whether we have the will to use that power. What distinguishes me from Johnson is that I have the will in spades." Indeed, Nixon vowed, "The bastards have never been bombed like they're going to be bombed this time."[23]

But the task would prove daunting. By May 1972, Hanoi had used the long bombing halt and the steady influx of imported SA-2 surface-to-air missiles, antiaircraft artillery (AAA), Soviet and Chinese MiG fighters, and sophisticated radar guidance technology to build one of the world's most robust air-defense systems. This was North

Vietnam's so-called "Defensive Triad." Communist anti-air tactics relied on both vertical and horizontal defense-in-depth, forcing U.S. strike pilots to run a murderously layered gauntlet to deliver their payloads.[24]

At the upper-most layer, the SA-2 lurked as the principal threat, especially to high-flying B-52s. Developed in the mid-1950s and known as the V-750 *Dvina* in the Communist Bloc, it was the first proven Soviet surface-to-air missile, shooting down American U-2 spy planes over the USSR and Cuba in 1960 and 1962. The Soviets began supplying the North Vietnamese with SAMs and operational training at the start of Rolling Thunder in spring 1965. The length of a telephone pole and weighing nearly 5,000 pounds, the SA-2 pulled Mach 3.5, had a ceiling of 60,000 feet, and an effective range of between 25 and 31 miles. And a direct hit was not necessary. A proximate shrapnel blast from its 288-pound fragmentation warhead could tear an enemy aircraft to bits. A typical SAM site consisted of six mobile launchers, support and technical crews, computer-assisted Spoon Rest detection radar, and a Fan Song radar guidance system. Spoon Rest could pick up enemy aircraft up to 70 miles out. From there, Fan Song took over target acquisition and guidance, tracking up to four aircraft simultaneously, and capable of guiding up to three missiles on target.[25]

The SA-2 presented several malign complications. First, it required U.S. planners to devote considerable resources to evasion and suppression. To be sure, strike missions would have been suicidal without an accompanying phalanx of escort fighters, F-4s to spread radar-confusing aluminum chaff corridors, SAM suppression aircraft like the F-105 Wild Weasels and A-7 Iron Hands, and radar- jammers such as the Douglas EB-66C, a USAF light bomber reborn as an electronic countermeasures (ECM) platform. Making matters worse, active SA-2 firing units were mobile and usually rotated daily among some 300 prepared SAM sites throughout the country. By spring 1972, North Vietnam possessed about 4,000 SA-2s and had even repositioned some launchers inside the DMZ and along the northwestern Laotian border with South Vietnam.[26]

The SA-2 was ineffective below 3,000 feet, but the missile threat forced tactical strike aircraft to lower altitudes where they then ran into the communists' second layer of air defense—AAA guns. By 1972, the North Vietnamese deployed about 5,000 of the weapons, many radar guided, ranging from 23 to 100 mm. Like the SA-2, AAA emplacements were often repositioned daily to maximize surprise and unexpected firing concentrations. If pilots managed to survive the larger antiaircraft guns, they were then met with a hail of heavy machinegun and small arms fire as they arrived over target. Together, AAA and smaller arms accounted for 77 percent of Air Force losses and 52 percent of Navy shootdowns, claiming more American aircraft than all other North Vietnamese air-defense assets combined.[27]

Finally, the third leg of the Triad often loomed large during American strike missions over North Vietnam. Aided by increasingly sophisticated tactics, a robust early warning radar system, and guided by ground control intercept (GCI) centers to help vector MiG pilots toward U.S. aircraft, North Vietnam's growing interceptor fleet posed a burgeoning threat. While MiG fighters could strike at any time, their pilots were especially keen to ambush American aircraft as they clawed for altitude following their attack runs. U.S. pilots were most vulnerable at that moment as they desperately sought to escape the torrent of AAA blistering from below. By spring 1972, the North Vietnamese arsenal including nearly 250 fighter aircraft dispersed among four fighter regiments, including 120 MiG-15/17s, 33 MiG-19s, and 93 of the vaunted MiG-21s, the Soviets' premier air interceptor at the time. To date, U.S. pilots had had relatively few encounters with either the MiG-19 or 21. That would soon change as American fliers were increasingly challenged by the 19's low- altitude dogfighting capabilities and the 21's vastly improved speed and sophisticated Atoll air-to-air missiles.[28]

Interestingly, the initial success of the North Vietnamese invasion had forced U.S. planners into action even before Nixon's decision. The Air Force and Navy had begun moving additional assets in theater the previous month, with a few Marine and Air Force F-4 Phantoms arriving from bases in Japan and Korea in early April. By 11 April and the advent of Operation Constant Guard, things ramped up considerably, as the first of more than 100 F-4s, along with two squadrons of C-130 transports began arriving from the United States. Meanwhile the Air Force's Operation Bullet Shot, begun as a trickle in February, had by April become a flood as the Strategic Air Command—whose primary responsibility was to act as one-third of the nuclear triad—moved mammoth B-52s and KC-135 refueling tankers into theater. Before long, airfields at U-Tapao Thailand and Guam's Andersen swelled to overflowing with nearly 150 of the heavy bombers. Gen. John Vogt, a World War II fighter ace who had recently assumed command of the Seventh Air Force, was the man responsible for directing Air Force operations in Southeast Asia. He would soon put the newly arrived assets to devasting use. For its part, naval Task Force 77 under Rear Adm. Damon Cooper upped its strike complement to six carrier air wings and some 500 aircraft by mid-May. In less than a month, the U.S. had more than doubled the strike aircraft on hand, with more on the way. And U.S. pilots would have new and effective ways to exploit them.[29]

On the air-to-air front, several innovations had developed since the end of Rolling Thunder. The Navy's Top Gun program, along with the Air Force's Fighter Weapons School, had helped pilots improve their dogfighting skills to counter the Soviet-style air tactics and more maneuverable MiG fighters employed by the Vietnam People's Air Force (VNPA). Meanwhile, the new top secret APX-80 system was a potential game-changer. Codenamed "Combat Tree," the system was equipped in a limited number of Air Force F- 4D Phantom IIs and allowed the back seat weapons systems officer (WSO) to identify enemy aircraft from beyond visual range. VNPA MiG-21s and even some variants of the MiG-19 were outfitted with an Identification Friend or Foe (IFF) transponder beacon to announce it as a "friendly" aircraft to ground gunners. This helped head off accidental targeting by North Vietnamese ground-controlled interception (GCI) stations and SAM radar systems. APX-80 turned the system on its head. Combat Tree could discreetly scan airspace within a 60-mile radius and at any altitude to interrogate transponders it found—all without requiring the WSO to switch on his aircraft's own powerful radar, thus alerting the MiG pilot he was being scanned. Enemy IFFs were designed to automatically respond with a code—again without the MiG pilot's knowledge—which the WSO could use to identify it as hostile. Combat Tree helped U.S. pilots improve their kill ratio, as well. Since most aircraft over Vietnam were American or South Vietnamese, and standard radar could detect only their presence, not identity, airmen had previously been required to close to visual range before firing. Now, fighter jocks equipped with Combat Tree could use their AIM-7E Sparrow radar guided missiles to target enemy aircraft over the horizon, regaining the element of surprise while better avoiding close-in dogfighting duels with the more maneuverable MiGs.[30]

Ground attack pilots had a few new weapons in their arsenal, as well. During Operation Rolling Thunder, air crews had to rely mostly on so-called "dumb" bombs—unguided, free fall munitions— to hit high-value military targets like bridges, power plants, rail depots, communication sites, etc. The limitations of this approach were many.

The general inaccuracy of such munitions—just 50 percent of which could be counted on to hit within 420 feet of the target—required a high number of bombs to have any hope of damaging, let alone destroying, a given target. This, in turn, required greater numbers of strike and support aircraft to deliver them. It also obliged pilots to release payloads from lower altitudes and from steeper dive angles to be effective. Both factors created a target-rich environment for North Vietnam's antiaircraft gunners. More than 900 U.S. planes would be shot down by the end of Rolling Thunder. Finally, despite pilots' best efforts, the use of unguided munitions— especially in a tactical environment where high value military targets were often located near civilian population and infrastructure— sometimes inflicted collateral damage that inflamed global and domestic antiwar sentiment. According to official U.S. estimates, some 30,000 civilians were killed during Rolling Thunder, with North Vietnamese claims running much higher.[31]

By spring 1972, however, several "smart" weapons had emerged. The Navy's AGM-62 Walleye, a free-fall munition originally developed as an air-to-surface missile, used a television tracking system to lock onto its target. While the camera system was crude and the munition performed poorly in inclement weather, the Walleye had some success toward the end of Rolling Thunder. Air Force research and development picked up the torch, eventually producing two early smart munitions: the Homing Bomb System (HOBOS) and the Paveway laser guided bomb (LGB) system. The former featured electro-optically guided bombs (EOGB) directed by an improved television tracking system that homed on the image contrast between the target and its background. Once released, the autonomous EOGB self-directed to target. The latter required Paveway F-4s to operate in pairs. The first Phantom was equipped with the Pave Knife laser designator pod used to continuously paint the target with laser light.

Meanwhile, bombs carried by the second F-4 featured a laser-seeker and guidance control unit in the nosecone. Once the target was illuminated, the Phantom pilot released his LGB—often the 2,000-pound MK-84— which homed on the reflected energy. The Paveway system eventually proved accurate to within 20 feet, greatly improving strike sortie effectiveness while requiring fewer bombs, fewer aircraft to deliver them, and greater pilot safety by allowing bomb release from much higher altitudes. Still, only seven F-4s would be equipped with the Pave Knife pod, so the system would be reserved for only the most dangerous missions, usually against hard targets in and around Hanoi and Haiphong. By early 1973, the Air Force would drop some 10,500 LGBs, with half scoring direct hits and another 4,000 landing within 25 feet of the target, making them some 100 to 200 times more effective than conventional bombs, according to one assessment. While the Air Force led the way in developing laser-guided munitions, the Navy soon wanted in on the action. By the latter half of 1972, the service had begun testing LGBs under combat conditions over North Vietnam.[32]

One of the pilots who volunteered to put the new weapon through its paces was Lt. Gerald "Jerry" Tucker, an F-8J Crusader pilot with VF-211 aboard the USS *Hancock.* A designated F-8 squadron, the Fighting Checkmates had already made a name for themselves, becoming unofficially known as the "MiG Killers." The Crusader would earn the best kill ratio of the war, downing 18 MiGs against just three losses, and the Checkmates alone would account for eight of the shootdowns. Many believed the fast and maneuverable F-8—nicknamed the "last of the gunfighters" because its quad-20-mm cannons were its primary weapon—played a prominent role in its own success. There was certainly something to the claim. On 21 June 1972, Lt. Richard "Dick" Evert with *Hancock*'s VF-24 Red Checkertails used his Crusader to scare off a pair of MiGs as they closed on a downed F-4 crew bobbing helplessly in the Tonkin Gulf. Evert's wingman had lost his radio, so the pilot took on both enemy fighters alone. When the North Vietnamese spotted the lone Crusader burning inbound, both MiG pilots turned tail and fled. "The aircraft was extremely maneuverable, a very favorable thrust-to-weight ratio and could climb in the vertical," says Evert. "We could outclimb the MiGs, could get above them, and when they fell away because of their lower thrust-to-weight ratio, we would be able to attack them from behind."[33] As for Tucker, he had the distinction of scoring perhaps the most unique MiG "kill" of the war. While flying target combat air patrol (TARCAP) for a strike mission near Haiphong on 23 May 1972, Tucker got into a near-fight with a MiG-17 that tried to jump the strike element on its way out. It was a "near fight" because the MiG pilot voluntarily punched out rather than face Tucker and his Crusader. Because he had not used weapons to make his "kill," the Navy declined to credit Tucker, but the exploit became legend among his fellow aviators.[34]

In late summer, the Navy delivered several LGBs from its southern California test range at China Lake to the *Hancock* for a series of trial runs. But the missions would be dangerous. The designator plane, usually an F-4, would have to orbit below 10,000 feet and slower than 250 knots to paint the target—a juicy "grape" for SAM and AAA crews. The back-seater had to ignore all the chaos around him as he gazed through a handheld scope to keep the crosshairs on target. Once the aimpoint was steadily illuminated, an attacking A-4 released the LGB into the resulting laser cone. His tour winding to a close, and with few other takers, Tucker stepped up. "Lots of guys said, 'I ain't doing that,'" Tucker says with a chuckle. "I was getting ready to leave anyway, so it sounded like fun to me." Although Tucker's Navy career would stretch more than 20 years— including a stint flying with the famed Blue Angels—the LGB missions would remain some of his most indelible memories. "Absolutely amazing," he says. "I'm holding the crosshairs right on target. All of a sudden, you see this little streak coming out of the corner and then a lot of smoke. And when the smoke cleared, no more target. It completely wiped it out. Hit exactly where we were looking. It was just phenomenal."[35]

These and other weapons would soon be put to devastating effect. Although Le Duan's audacious conventional offensive had shaken Saigon and Washington to their very core, his forces were now massed and in the open—a target-rich environment for U.S. and RVNAF airpower. Tactical close air support, guided by U.S. advisers on the ground and American forward air controllers (FACs) prowling the skies above, laid into North Vietnamese troop formations with merciless efficiency. Meanwhile, the Air Force nearly tripled its B- 52 Arc Light strikes to 2,000 by the end of May. The sorties employed three-plane cells to saturate preset kill boxes with up to 90 tons of bombs at a time, rumbling the earth and sending thunderous shockwaves miles in every direction. The effect was as physically and psychologically devastating to NVA troops as it was uplifting to South Vietnamese morale. Entire communist units simply ceased to exit, with few safe areas to the rear and lines of supply and communication constantly imperiled by air attack.[36]

While hammering invading communist forces in South Vietnam continued to take priority in April and early May, a few limited air operations had already begun to have an impact above the 17th parallel, as well. On 6 April, Operation Freedom Train commenced, targeting the North's logistics and military infrastructure, along with supply and communication lines stretching from the 20th parallel south to the DMZ. Designed to strangle the Nguyen Hue Offensive by cutting the supply of men and materiel flowing south, the strikes were the first against central North Vietnam since the end of Rolling Thunder. B-52 Stratofortress crews were also getting a piece of the action, as attacks on 9 and 12 April hit petroleum facilities at Vinh and the MiG airfield at Bai Thuong north of Thanh Hoa. The strikes marked the first use of heavy bombers over the North since October 1968.[37]

But using the big jets over North Vietnam's heavily defended heartland entailed significant risk. With its six man crew, the $10 million aircraft ($75 million in 2024 dollars) usually delivered its ordnance from 30,000 feet or above, so danger from the thousands of antiaircraft guns ringing the country was virtually nonexistent. Likewise, while MiGs could climb to meet the bombers, fighter escorts were usually able to keep them at bay. B-52 tail gunners would, however, end up bagging two of the enemy fighters by the end of the year. No, says then-Sgt. Mark Tiedemann, who logged more than 700 combat hours behind the B-52D's quad-50 caliber turret, it was the SA-2 that worried crews the most. From his aft bubble canopy, the 307th Strategic Wing tail gunner had spied more SAM contrails than he cared to count. But the day he saw one coming for *his* plane raised the "pucker factor" to a whole new level. "There was indeed an 'oh shit' moment when one of those delightful telephone poles turned into a black dot with a ring of angry fire," he says. As he watched the missile loom larger, Tiedemann keyed his headset. "Pilot, break left," he said as calmly as he could. But the big bomber had moved into the "box," that point of no return where any evasive action would too badly disrupt its guidance gyros to have any hope of hitting the target. "Sorry, guns, can't do it," came the reply. "We're too close to release." And then suddenly, the missile simply exploded harmlessly. A chaff cloud had likely triggered the SAM's proximity fuse prematurely. But Tiedemann, who would go on to pastor his own church following a 34-year Air Force career, has a different take. "I like to think God reached out His finger and put out that candle," he says with chuckle. Over the course of the war, the United States would lose 17 B-52s in combat, all to the Soviet SA-2 *Dvina*.[38]

Nevertheless, Nixon had no interest in nibbling at the edges this time. He wanted to strike right at the heart of North Vietnamese power and industry. His senior advisers were by no means united on the prospect, however. While Deputy National Security Advisor Alexander Haig and the Joint Chiefs were bullish on the idea, Kissinger and Secretary of Defense Melvin Laird were much more skeptical. Kissinger worried that hitting the North Vietnamese heartland too hard would derail the upcoming Moscow summit—and the continued prospect of détente—while Laird was concerned that the intensified bombing would prompt Congress to finally cut off funding for the war. A compromise was reached with the limited, one-day campaign dubbed Operation Freedom Porch Bravo. On 16 April, some 17 Thailand-based B-52s, supported by scores of Air Force and Navy tactical aircraft, hit key petroleum-storage facilities near Hanoi and Haiphong, dropping some 460 tons of bombs from 32,000 feet. Second and third waves of more than 100 tactical strike planes hit 10 more targets later in the day. MIGCAP fighters also shot down three responding MiG-21s, a fourth crashing upon landing. Another three MiG-17s were caught on the ground at Kien An airbase. Ultimately, the single-day operation knocked out about half the North's petroleum, oil, and lubricants (POL) stockpiles, along with large stocks of other military supplies. SA-2 missiles shot down one F-105G Wild Weasel and one A-7E.[39]

Eleven days later, Nixon once more unleased the dogs as the Air Force's 8th Tactical Fighter Wing out of Ubon Thailand hoped to score an edifying victory by at last knocking out Thanh Hoa's infamous Dragon's Jaw bridge. Traversing the Song Ma river and supporting the only railroad in the North Vietnamese panhandle, the double-spanned, 540-foot-long Dragon's Jaw was utterly indispensable to the war effort in the South. The imposing structure of steel truss reinforced concrete—40 feet thick in places—had withstood nearly 900 attack sorties since 1965, claiming 11 U.S. aircraft in the process. Heavy cloud cover precluded the use of Pave Knife laser designators, so Phantom pilots were forced to rely on 2,000 pound electro-optically guided bombs to do the deed. Despite damaging some highway sections, however, the bombs had little effect. Air Force F-4s would return on 13 May armed with laser guided bombs, including 3,000 pounders. Under better weather conditions, the LGBs found their mark, blasting the Dragon off its concrete abutments and knocking it out for the duration of the war. It was an early and dramatic statement on the power and precision of these new wonder weapons.[40]

Surprisingly, these intensified airstrikes drew little more than a routine complaint from the Soviets. It seemed General Secretary Leonid Brezhnev had no intention of letting Southeast Asian affairs hamper the Moscow summit. And besides, the Soviets were not about to back away from détente with the United States after Nixon's historic visit to rival China just two months earlier. As for the PRC, Beijing's own muted response signaled that it too was willing to overlook the stepped-up air campaign to keep improved U.S.-Sino relations on track. Especially appealing had been Nixon's momentous concession over the Taiwan issue during his February visit. In the so-called "Shanghai Communique" that emerged, the PRC reiterated its longtime assertion that Taiwan was not an independent country but merely a renegade Chinese province that must eventually be "reunited" with the mainland. In the document, the United States for the first time acknowledged "that all Chinese on either side of the Taiwan Strait maintain there is but one China and that Taiwan is a part of China…[and] reaffirms its interest in a peaceful settlement of the Taiwan question by the Chinese themselves." It was an astonishing about-face for U.S.-Taiwan policy. Nixon's concession would pave the way for the United States' formal recognition of the Chinese Communist Party as the sole government of China in 1979—while simultaneously breaking off diplomatic relations with Washington's longtime allies in Taipei. The resulting "One

China" policy continues to color U.S., Chinese, and Taiwanese relations to this day. Nevertheless, it appeared Nixon would have a relatively free hand over North Vietnam in the short term.[41]

On 9 May, almost simultaneously with the president's televised address to the nation, the U.S. Navy commenced Operation Pocket Money to mine North Vietnamese ports and inlets. Foremost among them was Haiphong Harbor and its approaches, especially its Cua Nam Thien channel, a narrow, mile-long passage that admitted access to the harbor itself. Aided by diversionary strikes on Thanh Hoa and Phu Qui by aircraft from the USS *Kittyhawk*, along with coastal shelling by Navy destroyers, six A-7E Corsairs and three A- 6A Intruders under the command of Cmdr. Roger Sheets launched from the USS *Coral Sea* to sow the harbor's main shipping channels with 36 MK-52 electromagnetic mines. Despite abundant antiaircraft assets in the area—the pilots came in fast and low to avoid SAM radar—the mission was a complete success, with all aircraft delivering their payloads in a matter of minutes before returning unscathed. The mission was not without excitement, however. At one point, radar detected three MiGs from Phuc Yen airbase streaking toward the *Coral Sea* strike force. The danger to surface ships was real. A month earlier, two MiG-17s had attacked the destroyer *Higbee,* a 500-pound bomb knocking out the ship's aft five-inch gun and wounding four sailors. The nearby light cruiser USS *Oklahoma City* also sustained minor damage. To prevent a replay, the guided missile cruisers *Chicago* and *Long Beach* had been positioned closer to shore to screen against enemy fighters. Despite coming under fire from North Vietnamese coastal batteries, *Chicago* let loose with two Talos surface-to-air missiles, downing one MiG and scattering the others. Kinetic action, however, was not *Chicago's* chief responsibility. She was one of several cruisers that since 1965 had served as the Navy's Positive Identification Radar Advisory Zone (PIRAZ) station in the Tonkin Gulf. Call sign "Red Crown," the

PIRAZ vessel operated akin to a radar picket, tracking enemy aircraft over eastern North Vietnam and the Gulf. Equipped with powerful radar, computing, and communications gear, Red Crown's radarmen also became indispensable in directing air-to-air missions. *Chicago's* Senior Chief radarman Larry Nowell would that August earn the Navy's Distinguished Service Medal—a rare honor for an enlisted man—for helping U.S. pilots destroy 12 North Vietnamese MiGs. The Air Force operated similar platforms codenamed "Disco," and later "Teaball." But due to various technical challenges, including the need for a radio relay via an orbiting EC-121 Warning Star, Air Force crews came to rely upon Red Crown almost as much as their Navy counterparts.[42]

Pocket Money mines were set to activate within 72 hours, and all foreign-flagged ships were warned to leave. Only five vessels would take heed, however, leaving 31 foreign merchant ships trapped in Haiphong Harbor. These included 10 Soviet-registered and five Chinese ships, with another half dozen hailing from Eastern Bloc countries. Two other Soviet cargo ships were trapped at nearby Cam Pha, with another two Chinese vessels bottled in the port of Vinh 280 miles south. Navy and Marine mining operations against nearly all North Vietnamese ports and inlets would continue over the next eight months. In all, nearly 12,000 mines would be laid. Of course, the port city of Sihanoukville, in supposedly "neutral" Cambodia, remained off limits. As they had from the start, Chinese and Soviet vessels throughout 1972 continued to disgorge massive quantities of war materiel at its docks. Supplies were then ported along the so-called "Sihanouk Trail" to staging areas opposite South Vietnam's western borders. The volume of communist war goods flowing via the Cambodian port rivaled and, at times, even exceeded that of the vaunted Ho Chi Minh Trail. Still, the mining effort, along with a concerted naval blockade of the North Vietnamese coast, would severely hamper seaborne imports, forcing the bulk to travel by rail and truck. While foreign merchant ships were off limits, no such restrictions hampered allied air attacks on overland transport. This witches' brew was meant to strangle Hanoi's ability to resupply both its war in the South and replenish badly needed anti-aircraft assets in the North.[43]

And this was just the tip of the iceberg. Nixon charged Moorer with drawing up a comprehensive air campaign to accomplish the mission he had laid out. A noted football enthusiast, the president had codenamed himself "Quarterback" during Kissinger's secret peace negotiations in Paris. It was perhaps fitting then that the new bombing campaign be dubbed "Operation Linebacker." For airstrike planning, North Vietnam had long been divided into seven regions known as "route packages." The Seventh Air Force was responsible for all targets from the DMZ up to the 18th parallel (RP 1), along with RPs 5 and 6A encompassing northwest and north-central North Vietnam, including Hanoi. Task Force 77 was charged with targeting Haiphong and the rest of North Vietnam's central and coastal regions designated RPs 2, 3, 4, and 6B. Set to begin on 10 May, Operation Linebacker would prove to be the most comprehensive and devastating application of airpower of the war. Still, this would be no indiscriminate bombing campaign. Moorer insisted that everything possible be done to avoid civilian casualties, and warned commanders they would be held personally responsible if that directive were violated.[44]

And so, the die was cast. But serious questions remained. Would Linebacker accomplish Nixon's dual objectives of crushing the North's invasion while forcing Hanoi back to the bargaining table? And what of the Soviets and Chinese? Would this sustained and intensive bombing campaign doom the Cold War thaw after all? Back home, what domestic repercussions lie in wait? Despite a recent Gallup poll showing that 74 percent of those interviewed supported Nixon's harder line against North Vietnam, would the always-simmering antiwar movement explode once again, at last giving Congress the final excuse to cut off funding for the war? Lastly, might all this end up costing Nixon not just a second term in office, but even his pledge of an "honorable peace" in Vietnam? The coming months would tell the tale.[45]

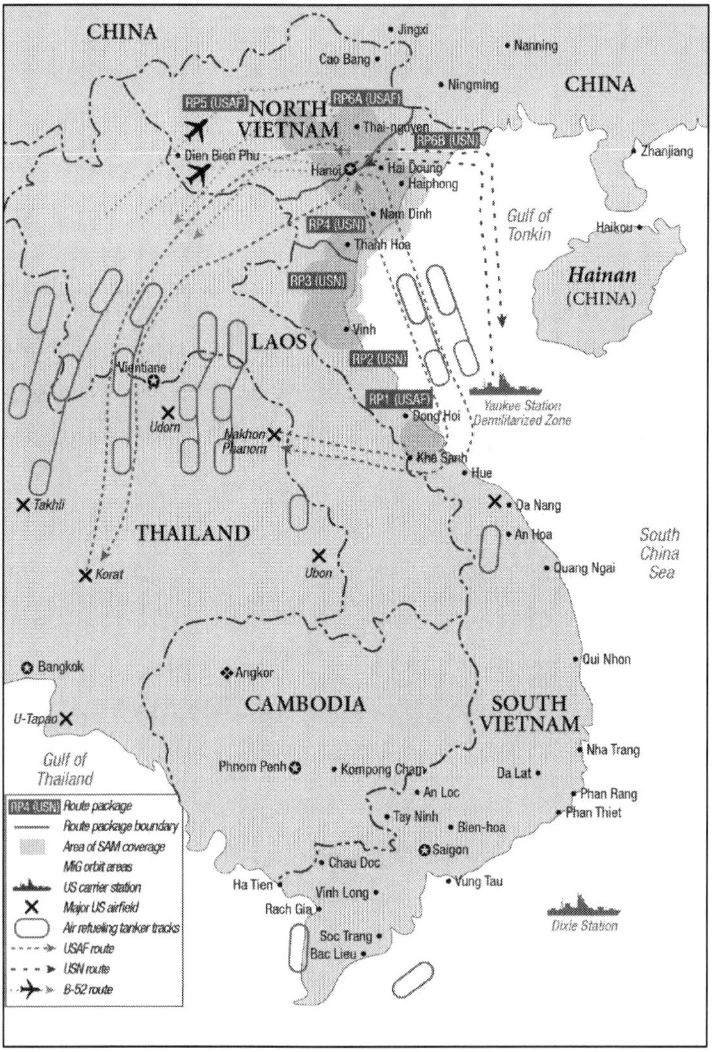

Figure 2: Operation Linebacker

4

LORDS OF AZURE

10 May

Ironically, the first blows of what would become the war's most intense air campaign were dealt not by air but by sea. Just after 0200 hours on 10 May, Seventh Fleet commander Vice Adm. William Mack's flotilla of four cruisers and two destroyers moved into position off the coast near Haiphong. The formation was part of Task Unit 77.1, the Seventh Fleet's surface component charged with bombarding North Vietnamese coastal installations, interdicting supply vessels, and providing gun support for allied land operations. With her sister ships tasked with providing covering fire, the cruiser *Newport News* let loose with her eight-inch guns, hurling 335-pound shells some 14 miles toward targets along the coast. Thanks to the U.S. bombing halt, along with years of unfettered materiel imports from the Eastern Bloc, North Vietnam's coastal defenses—like its air defense network—had developed into one of the world's most formidable. Dozens of shore batteries nestled around Haiphong alone. And now those big guns answered in a thunderous display of fire and fury. On cue, the rest of Mack's task force unleashed a blistering counter barrage, incoming and outgoing artillery lighting the night in blinding yellow-orange flashes. In minutes it was all over, as Mack's flotilla faded back into the gloom unscathed. But the *Newport News* had done her work, ripping off nearly 120 eight and five-inch shells, targets ablaze all around Haiphong. But the opening salvo had not been without risk. While the North never managed to sink a U.S. vessel during the war, gun battles with its coastal defenses routinely left Navy ships severely damaged, often with dead and wounded sailors aboard. Operation Linebacker would be no different. By December, Task Unit 77.1 would hit targets all along enemy shores—including Haiphong five times—sending more than 111,000 rounds to knock out coastal guns, supply convoys, and logistics networks, while sinking some 200 North Vietnamese surface vessels.[1]

And the day was just getting started. Steaming north from Yankee Station near the DMZ were three U.S. carrier groups: the *Constellation* (CVA-64), *Kitty Hawk* (CVA-63), and *Coral Sea* (CVA-43), each with a capacity of about 90 fixed wing aircraft. Haiphong was again in the crosshairs, its main POL storage facility at the top of the target set. This was the same objective that Air Force B-52s had badly damaged during Operation Freedom Porch Bravo in April, and *Connie's* air crews were intent on finishing it off once and for all. Cmdr. Gus Eggert, commanding *Constellation's* Carrier Air Wing 9 (CVW-9), ordered the first wave of 33 strike and support aircraft off the deck at about 0730 hours. Refueling tankers, including the A-3 Skywarrior, orbited at 15,000 feet to top off the strike force before it banked northwest toward the port city. Nicknamed the "Whale" because of its prodigious girth, the A-3 was originally designed as a nuclear bomber but had since been repurposed as an aerial tanker capable of hauling nearly 24,000 pounds of fuel. It was the largest aircraft capable of carrier launch and had become the Navy's most effective tanker of the war. A single E-2B Hawkeye, launched earlier to give the slow-flying prop plane a head start, orbited just off the coast, its massive radar dish and sophisticated scanning and communication equipment acting as the strike force's early warning eyes and ears.[2]

As the strike group neared the target, two SAM-suppressing A-

7 Corsairs swept ahead prowling for targets. Codenamed Iron Hand because its pilots needed "nerves of steel" to fly into the teeth of anti-aircraft defenses, the program was a joint Navy-Air Force effort begun in 1965 for the suppression of enemy air defenses (SEAD). Together with their Air Force counterparts, the Wild Weasels, Iron Hands were equipped with advanced avionics, cluster munitions, and anti-radiation AGM-45 Shrike missiles to home in and hamper or destroy SA-2 and other radar-guided anti-aircraft defenses.[3]

Next came F-4 flak suppressors armed with Rockeye cluster bombs to hit the unguided antiaircraft gun emplacements ringing the target. Initially designed as a tank killer, the Rockeye had proven particularly effective against antiaircraft positions, as well. Another eight Phantom IIs would screen north, west, and south of the city to deal with enemy fighters from nearby MiG bases. Finally, 17 A-6s and A-7s would pack the punch, each carrying between 6,000 and 8,000 pounds of MK-82 bombs. Similarly comprised attack formations launched in successive waves from *Coral Sea* and *Kitty Hawk*, each bound for its own target set in the Haiphong area. A handful of SH-3 Sea King choppers from the USS *Okinawa* stood by for combat search and rescue (CSAR) 30 miles off the coast. Finally, an EKA-3B—part tanker, part electronic warfare platform—circled at 20,000 feet to help disrupt enemy radar.[4]

As the *Constellation's* strike group closed on target, the sky erupted with radar-guided AAA, fiery red and black bursts all around. But the explosions were off target. The EKA-3B's jamming equipment was having the desired effect. The Phantom II flak suppressors dove on the orange muzzle flashes below, saturating the gun sites with thousands of cluster bomblets in a dazzling display of fire and smoke. Moments later, the telltale whine of radar warning receivers warbled over the strike pilots' headsets. The SAMs' Spoon Rest acquisition radar had locked on. The North Vietnamese let loose with a half dozen "flying telephone poles," their white contrails crisscrossing the sky. But before the big missiles could find their mark, the Iron Hands pounced, AGM-45s blasting loose from their pylons and shrieking earthward. The radar-homing Shrikes often intimidated ground crews into shutting down their Fan Song guidance radar to keep from being destroyed themselves. SA-2s already in the air would immediately lose their targets and careen harmlessly away—at least harmlessly for American pilots. SAMs that missed their targets—which happened the majority of time—fell indiscriminately. Mounting a 288-pound warhead of high explosives and weighing up to two and a half tons, the errant missiles often did as much damage as an enemy bomb, wreaking destruction and untold civilian casualties when they crashed.[5]

Now the strike element came on. By twos the *Connie's* Intruders and Corsairs tilted at 45 degrees and dove like raptors. At 7,000 feet, calls of "pickle, pickle, pickle," echoed across the radio net as pilots released their MK-82s. Fifty tons of ordnance carpeted the facility in a flaming maelstrom. Moments later, Haiphong's main POL storage complex was a blasted wreck, thick oily smoke snaking skyward. For good measure, the Iron Hands then streaked west of Haiphong to chew up the MiG airfield at Kien An with their unused cluster munitions. Meanwhile, the *Coral Sea* strike force worked over Haiphong's railyard, while *Kitty Hawk's* group hit the main rail bridge leading out of the city, destroying one span and cratering the highway.[6]

The mission was a complete success: all targets hit and badly damaged without a single aircraft lost. A near-miss SAM did manage to pepper a *Constellation* RA-5C Vigilante with shrapnel as it streaked overhead to snap post-strike photos, but plane and pilot returned safely. And with that, all three strike groups banked southeast, called "feet wet" as they cleared the coast, and streaked toward waiting carriers.

The attack had taken mere minutes. Then one final prize. Two of *Constellation's* VF-92 F-4Js flying MIGCAP west of the city managed to bag a shiny new MiG-21 near Kep airbase. It would be the first of eight enemy fighters splashed by Naval aviators that day—exceeding by one the service's MiG tally for the previous four years. Navy had certainly gotten Linebacker off to a good start. Now it was Air Force's turn.[7]

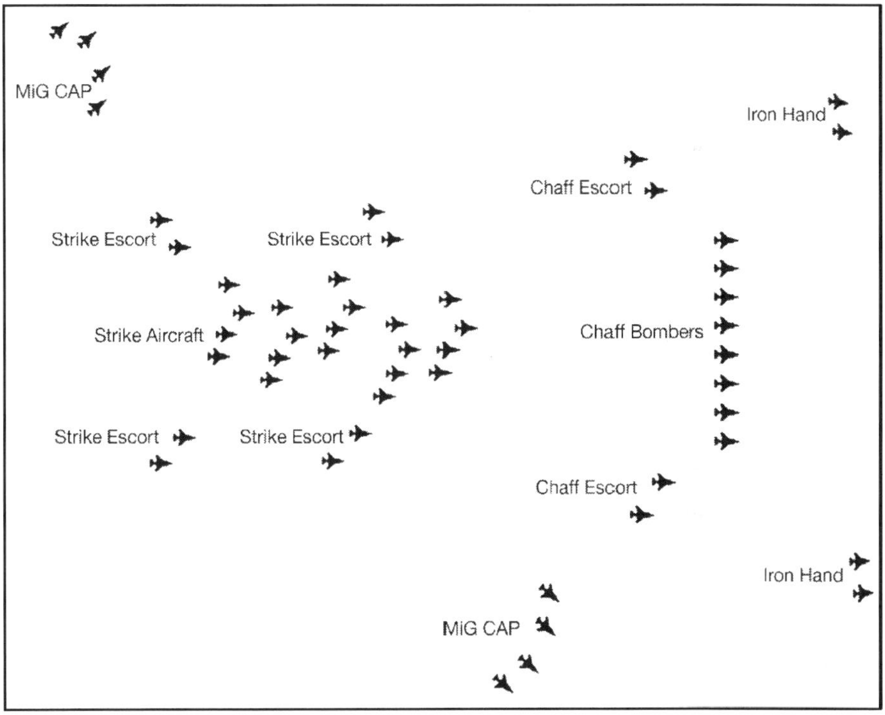

Figure 3: Linebacker Strike

Approximately 55 miles to the west, Maj. Robert Lodge and his four-ship Oyster Flight of the famed 555th Tactical Fighter Squadron (TFS)—the "Triple Nickle"—were running fighter screen for the Air Force's first Linebacker strikes against targets in downtown Hanoi. The service approached its missions somewhat differently than the Navy.

Where each aircraft carrier wing was completely self-contained and possessed all necessary assets to accomplish a given mission, various Air Force wings scattered among Royal Thai Air Force bases throughout neighboring Thailand focused on just one or two tasks each. For example, Udorn's 432nd Tactical Reconnaissance Wing (TRW)—of which the 555th was a part—would provide fighter escort and reconnaissance aircraft for the strike. Meanwhile, air wings based at Ubon, Korat, Nakon Phanom, and U-Tapao would each provide additional elements like attack craft, anti-aircraft defense suppression, radar jamming, search and rescue, and so on to round out mission requirements.[8]

The Hanoi strike force numbered some 117 attack and support aircraft, including F-4 bombers and escort fighters, F-105 Wild Weasels, chaff Phantoms, electronic warfare EB-66s, KC-135 fuel tankers, and more. As the various components assembled just beyond North Vietnamese airspace, Oyster Flight pushed into position northwest of the capital at about 0930 to lie in wait for any lurking MiGs. Lodge had with him two ships from another MIGCAP (Balter Flight) which had linked up with Oyster after two of Balter's F-4s were forced to turn back due to mechanical difficulties. Lodge, a skilled air-to-air tactician who, along with his back seat weapons system officer (WSO) Capt. Roger Locher, had already tallied two MiG kills, devised a trap. His Oyster Flight would maintain radio silence and come in fast and low—below 300 feet—to avoid detection. Meanwhile, the two Balter Phantoms would orbit at 22,000 feet as bait. When the MiGs moved to pounce on their "easy" prey, Oyster would roar up from the deck and spring the ambush. They did not have long to wait. At about 0945, Oyster received word from Red Crown's Chief Radarman Larry Nowell that there were four bandits burning from the north at 15,000 feet, the lead just 16 miles out, and all closing fast. These were MiG-21s of the elite 921st Fighter Regiment out of Kep airbase. Lodge and company immediately jettisoned their external tanks, hit afterburner, and climbed to intercept. Overhead, the MiGs came on, still unaware of Oyster's presence. Element lead Capt. Steve Ritchie, call sign Oyster 3, had flown 95 missions as an F-4 "Fast FAC" out of Da Nang in 1968. A relentless student and practitioner of air-to-air combat, he had at just 26 gone on to become the youngest instructor at the Air Force's Fighter Weapons School at Nellis. In fact, his current flight leader Lodge had been among his most talented students. Now back after volunteering for one final combat tour, Ritchie flew one of three Oyster Flight Phantoms equipped with the top secret APX-80 "Combat Tree" system that day. Suddenly, the MiGs flashed across the F-4's radar. His WSO Capt. Charles "Chuck" DeBellevue confirmed the bandits hostile as Combat Tree lit up with the MiGs' IFF transponder beacon. "He's squawking MiG…he's squawking MiG!" DeBellevue called. "Stand by to shoot!"[9]

Locher got confirmation the same moment, so Lodge pulled his Phantom into a climb and achieved full-system radar lock. The onrushing MiGs were still eight miles out and beyond visual range. Lodge squeezed the trigger on the stick, and the radar-guided AIM-7 Sparrow blasted free of its pylon. But it was a dud and exploded soon after launch. Lodge let loose again, the second 12-foot-long missile screaming to target. Meanwhile, his wingman Lt. John Markle and back seater Capt. Stephen Eaves in Oyster 2, just 600 yards off Lodge's right, squeezed off a pair of AIM-7s. Ritchie, 3,000 yards to Lodge's left, fired as well, but the Sparrow's motor failed to ignite, and simply fell away like a stone. Such problems were all-too- common with the AIM-7 in Vietnam. Haunted by design and reliability problems, Sparrows recorded fewer than one kill for every 10 missiles fired. Still, two of the five AIM-7s had found their mark, as both Lodge and Markle scored hits, brilliant yellow-orange fireballs erupting in the distance. Undaunted, the two remaining MiGs roared past. Lodge and Ritchie pulled their Phantoms hard in pursuit, eager to achieve rear-quarter positions on the North Vietnamese. At 6,000 feet, Ritchie squeezed off two more Sparrows, one detonating beneath the MiG's fuselage, its pilot ejecting as his aircraft burst into flames. "Oyster 3's a splash!" cried DeBellevue as the F-4 screeched past the MiG pilot's chute. It was the pair's first kill of the war—and the flight's third of the day[10]

Now Lodge locked in on what he hoped would be his fourth kill of the war. Oyster 1 rolled out of his hard turn and found himself trailing the MiG by just 200 feet. Too close for missiles, and his F- 4D had no guns that day. Lodge eased off the throttle. The range increased. *Just a few more seconds*. Suddenly, the pilot's headset crackled with a frantic warning from wingman Markle: "You got a bandit in your ten o'clock, Bob, level! Reverse right, reverse right, Bob! The bandit's behind you!" While Lodge and Locher had been zeroed in on the fourth MiG, a pair of Chinese-built Shenyang J-6s with the 925th Fighter Regiment out of nearby Yen Bai had roared up from below to spring their own ambush. Slower but more nimble than the MiG- 21, the J-6 boasted a trio 30 mm autocannons, ideal for close-in dogfighting. "Oyster 1 padlocked," Lodge called, then let loose with a Sparrow. But the J-6 was already blasting away. *Whump whump whump,* went the big guns, the F-4 shuddering under the weight of the heavy rounds. For a moment, Locher was sure they had collided with the fleeing MiG. But then he spied the fighter roaring away unscathed, exhaust plume shrinking in the distance. "Oh shit!" he exclaimed, still unaware of his peril. "The guy in front was getting away." More thuds as the J-6's 30 mms once again found their mark. The F-4 suddenly hemorrhaged airspeed, then was seized by a series of violent yaws, its right engine hit and burning wildly. Lodge, now aware he had gone from hunter to hunted in mere moments, announced all hydraulics were gone. The plane would soon drop like a stone. Flames scorched and bubbled the canopy as the engine fire consumed more and more of the aft section.

Now the WSO's altimeter showed that they had plunged below 8,000 feet. Time to go. "Hey, Bob…it's getting awful hot back here. I'm going to have to get out." As weapons officer for the 432nd TRW, Lodge was privy to sensitive information he believed invaluable to the enemy. He had vowed never to be taken prisoner and often said that should his plane be hit, he would not bail out. Lodge looked over his shoulder. "Why don't you eject then?" Then the plane rolled inverted, the earth where the sky should be. Now under negative Gs and pinned against the canopy, Locher clawed at the ejection handle between his legs and pulled with all his might. A split second later he blasted free of the stricken craft, rocketing headfirst toward the rolling countryside below. Miraculously, the chute deployed and yanked him upright just in time to see the pursuing J-6s roar past. A few moments later, Oyster 1 exploded in a massive fireball as the Phantom impacted the ground. Locher scanned for his friend's parachute but there was nothing.[11]

The WSO came down hard in a thickly wooded copse, bumping and crashing through the trees, some 40 feet tall. After he had worked his way down, Locher fought to free his billowing chute from the tangled limbs, but it was no use. The green, orange, and white panels of the draped silk stood out like a homing beacon for the enemy, who even now must be on their way. Still shell-shocked from the violent ejection, Locher could not seem to make his hands work well enough to free his large survival pack. After some minutes of fruitless fumbling, he abandoned the precious food and water it contained. Luckily, his smaller survival vest had a few invaluable items, including two pints of water, a knife and pistol, first aid kit, and mosquito netting and repellent. It would have to do. He pulled out his URC-64 rescue radio. "This is Oyster zero one Bravo, I'm on the ground. I'm ok." Then, fearing his transmissions would draw unwanted attention, he switched off. Besides, Locher knew he was far too deep inside enemy territory to chance a rescue. And he was right. Little did he know, but he had landed just five miles from the major air base at Yen Bai, itself some 95 miles northwest of downtown Hanoi. Regardless, he knew he had to get out of there— and fast. The enemy would scour the area in a desperate bid to capture the downed flyer. Locher hid his flight helmet and harness and then stumbled away to put as much distance as possible between him and his brightly colored parachute. He had made perhaps half a mile when he heard the urgent chattering of the North Vietnamese wafting from the direction he had come. They had already found the chute. And if he were not very careful—and very lucky—they would soon find him, too.[12]

Above, the remainder of Oyster Flight had watched in horror as Lodge and Locher's Phantom slammed into the earth. Not one among them had seen a chute, and all feared the worst. Stunned at the loss and fearing another MiG ambush, they decided the time had come to leave North Vietnam. "We had been sandwiched," Ritchie would later say. "We were lucky to come out of the engagement with only one loss." Standard operating procedure called for a fast and low egress. The Phantoms dropped to the deck and went supersonic, 850 miles per hour the absolute top end of the F-4's low-altitude speed. Oyster 2 had a head start and pulled away, leaving 3 and 4 bunched behind. Suddenly, Oyster 3 WSO DeBellevue realized they had company. A MiG-21MF was hot on their tail—and gaining fast. This was the top-line export variant of the Soviet's MiG-21. NATO codenamed it the "Fishbed-J." Whatever its name, DeBellevue was stunned the plane could move that fast, that low. Coming up close behind, the enemy fighter essentially "joined" the formation at one point. Just 300 feet off the left quarter position and about 100 feet above, the MiG had an ideal firing angle on both F-4s. And he had guns. Anxious moments passed. Then, for reasons known only to him, the MiG pilot stayed his hand, banked left, and roared toward the horizon. Oyster Flight streaked for home.[13]

Maj. Robert Lodge was never seen again. Originally listed as missing in action, he was officially declared dead on 9 May 1973. Over two tours in Southeast Asia, Lodge flew 186 combat missions, shooting down three North Vietnamese MiGs in the process. He was awarded five silver stars for gallantry in the face of the enemy. In September 1977, the Vietnamese government located Lodge's remains and eventually returned them to U.S. custody. As for Locher, he somehow managed to escape and evade his pursuers. Moving only in darkness and avoiding the many patrols and peasant farmers roaming the countryside, he subsisted on what few berries and natural forage he could find, slurping water from banana trees to survive. Finally, on 1 June, Locher recognized the telltale roar of U.S. aircraft overhead. He risked a radio call. "Any U.S. aircraft, if you read Oyster 1 Bravo, come up on Guard." But the airmen were suspicious, for there were no designated Oyster callsigns that day. The communist tactic of trying to lure and ambush rescue craft by impersonating downed American fliers was well known to all. Miraculously, none other than Capt. Steve Ritchie, his Oyster MIGCAP flight mate from 10 May, happened to be in one of the F- 4s. "My God," thought Ritchie. "That's Roger Locher!" After some back and forth, Locher finally deadpanned: "Guys, I've been down here a long time. Any chance of picking me up?" Within hours, rescue choppers, along with A-1H Skyraiders and F-4 fighter escorts, arrived to pull Locher from his hide. But MiG-21s soon appeared and, along with intense anti-aircraft ground fire, drove off the would- be rescuers. The WSO would spend another night in the wilderness.[14]

Word that Locher was alive and located rocketed up the chain to 7th Air Force commander Gen. John Vogt. Although Linebacker missions had now reached a fevered tempo, Vogt resolved to "shut down the war" and go all out to rescue Locher. "Goddamn it," he would later say, "the one thing that keeps our boys motivated is the certain belief that if they go down, we will do absolutely everything we can to get them out. I didn't ask anybody for permission. I just said, 'Go do it!'" Acting on his own authority, Vogt canceled all Hanoi strike missions for the following day. On 2 June, he ordered into action all available resources—some 150 aircraft, 119 of which would directly participate in the rescue attempt. Fighter-bombers and A-1Hs roared in to pound anti-aircraft positions all around Locher's location. Still, North Vietnamese anti-aircraft fire remained heavy and intense. Undaunted, USAF Capt. Dale Stovall and his HH-53C "Super Jolly Green Giant" rescue chopper braved the storm to retrieve the downed airman. Stovall, along with Skyraider pilot Capt. Ronald Smith, were later awarded the Air Force Cross for extreme heroism in helping to save Locher. The WSO's 23 days on the run, hunted and starving deep in enemy territory, would be the longest stretch endured by any downed airman during the war. Locher lost 30 pounds for his trouble—but kept his life. He returned to a hero's welcome at Udorn, where Vogt, who had flown up from "Blue Chip"—the 7th Air Force's Tactical Air Control Center in Saigon— stood by to welcome him. Locher would earn two Silver Stars for his service in Vietnam. Upon his return to the States, he trained for the pilot's seat and wound up flying F-4s and later F-16s for years to come. He retired a colonel in 1998.[15]

For captains Ritchie and DeBellevue, the events of 10 May marked only the start of their Linebacker exploits. In coming months, the pair would team for three more MiG kills, becoming the most prolific air-to-air combat duo of the war. On 28 August, Ritchie would earn his fifth kill (and DeBellevue his fourth), achieving the highly coveted status of "fighter ace." Fearing the target it would place on his back among enemy pilots, the Air Force then pulled Ritchie from the cockpit. He was awarded the Air Force Cross and four Silver Stars for his actions over Southeast Asia. After a 35-year career, Ritchie retired a brigadier general in February 1999. As for DeBellevue, his work was not yet done. Pairing with new pilot Capt. John A. Madden, the WSO went on to score his fifth and sixth kills while flying MIGCAP west of Hanoi on 9 September. His achievement would lead all services for the Vietnam War. He, too, was awarded the Air Force Cross, along with three Silver Stars, and retired a colonel in February 1998 after 30 years of service. The Air Force would achieve its third and final ace of the war when on 13 October WSO Capt. Jeffrey Feinstein downed his fifth enemy fighter while flying MIGCAP with his 13th TFS commander Lt. Curtis Westphal near Kep airfield northeast of Hanoi. Feinstein was also awarded the Air Force Cross along with four Silver Stars for his service in Southeast Asia. He retired a lieutenant colonel in July 1996. Feinstein remains America's most recent fighter ace.[16]

§

While Oyster Flight had been fighting its deadly duel, aircraft for the Hanoi strike mission were preparing for their attack runs. The targets were the notorious Paul Doumer Bridge just east of downtown and the vital railyard just north at Yen Vien. While the railyard was an important logistics and transportation hub for materiel arriving from China, the bridge had been an especially tempting—and frustrating—target for U.S. planners and airmen since 1967. Opened in 1903 and named after the then-governor general of French Indochina, it was renamed the Long Bien Bridge after the French defeat in 1954. Most Hanoi residents, however, continued to refer to it by its original appellation. Over a mile and a half long, and accommodating both rail and road traffic, the Paul Doumer spanned the Red River and provided Hanoi's only rail link with the crucial port city of Haiphong 55 miles east. Designed by architects Dayde & Pille of Paris, it was a massive steel truss structure bolstered by thick, reinforced concrete that proved remarkably resilient in the face of repeated U.S. air attacks. The bridge had been partially knocked out several times over the years, but North Vietnamese efforts to repair and reopen were ongoing. The bombing halt had allowed crews the respite to fully restore the bridge, which had remained in continuous operation ever since. Nevertheless, Hanoi's leaders, fully aware of the Doumer's importance, continued to pour in resources to bolster its defense. Numerous SA-2 launch sites, along with hundreds of anti-aircraft guns, some as large as 100 mm, ringed the structure. Three MiG bases, just minutes away by air, presented a lethal air-to- air threat, as well. As it had since the beginning, Paul Doumer would prove a deadly challenge for U.S. pilots.[17]

But the Air Force had an ace up its sleeve. Where previous sorties against the bridge had sometimes required as many as 60 strike aircraft and 100 tons of ordnance, new U.S. smart weapons would drastically lighten the load. U.S. planners now gauged that a mere 16-plane strike element armed with just 29 tons of precision- guided bombs could down the Doumer. Four F-4D/Es of the 435th Tactical Fighter Squadron (TFS) out of Ubon would carry two 2,000- pound electrooptically guided bombs (EOGB) each while another 12 Phantoms of Ubon's 8th Tactical Fighter Wing (TFW) would each load a pair of 2,000-pound laser guided bombs (LGB).[18]

At about 0940, the Doumer and Yen Vien strike groups came up from the south, Hanoi's air raid sirens wailing, the bridge strike force about five minutes ahead. The groups were to hit each target in rapid succession. To the west, four EB-66Es of Korat's 42nd Electronic Warfare Squadron (EWS) orbited at 30,000 feet just outside Hanoi's SAM umbrella to jam enemy radar. Eight chaff Phantoms out of Ubon led the way to further confuse AAA tracking. Each F-4 carried nine M-129 bombs packed with thousands of two- inch aluminum strips. These were to be dropped ahead and above of the oncoming strike group at

15-second intervals. The bomb casings were designed to burst open, allowing the strips to sprinkle and disperse snowflake-like to form radar-dizzying "chaff clouds." It was dangerous work. Like the World War II bomber formations of old, chaff Phantoms were required to fly a slow (no more than 575 miles per hour), undeviating course no matter what the enemy threw at them. Any evasive action to avoid AAA or even SAMs would cause the chaff cloud to improperly disperse, leaving the strike element highly vulnerable. Next came four F-105G Wild Weasels of Korat's 388th TFW roaring in to suppress nearby antiaircraft batteries and SA-2 sites with cluster munitions and radar-homing Shrikes. Finally, four F-4Es from the 336th and 432nd TFSs out of Udorn flew escort to deal with any nearby MiGs.[19]

North Vietnamese crews let loose with a hail of antiaircraft and surface-to-air missiles, the sky black with AAA, while SA-2s streaked here and there like deadly tree trunks. The chaff Phantoms, despite packing their own self-contained AlQ-87 radar-jamming pods, soon found themselves assailed by at least 42 of the big missiles. Down on the deck, the F-105s were in a ferocious fight of their own with the SAM sites. Two of the four Weasels, cockpits awash with the urgent squawking of radar warnings, quickly burned through all their anti-radar Shrikes in a desperate attempt to quell the firestorm. In minutes, the Weasels would fire off 25 AGMs in the effort. Still, only a series of skilled—and sometimes wildly lucky— maneuvers allowed the chaff F-4s and Weasels to survive unscathed. But now the chaff cloud had begun to do its work, swirling into an 18-mile-long corridor, two miles wide and a mile deep. As the shimmering metal strips began to play havoc with enemy guidance radar, the Phantoms and F-105s pulled up and away.[20]

Then the 16 F-4s of the strike group came on in waves at two-mile intervals, determined to blast the Doumer from its massive abutments. A second four-plane flight of F-105s, along with the four F-

4 fighter escorts, dropped low to lead the way. As before, the antiaircraft fire came up thick and ferocious, blotting the sky with angry bursts. The first wave, armed with 2,000-pound EOGBs, called "pickle" and released their ordnance at 12,000 feet. The results were disappointing. Despite their billing, not one of the eight optically guided bombs struck home, instead sending up great geysers of water but no concrete or steel. One bomb even vectored off wildly toward a nearby train station. The 12 F-4s equipped with K-84 laser-guided 2,000 pounders hoped to fare better. In successive waves, the three remaining four-ship flights dipped to 45 degrees and screeched earthward, letting loose at 12,000 feet. North Vietnamese gunners, many no doubt seasoned veterans of Rolling Thunder, blasted away but could not land a blow. The LGBs had enabled the pilots to release at a much higher—and safer—altitude than in years past. As the K-84s fell away, the WSOs trained their laser designators on the Doumer and watched with satisfaction as fiery smoke, concrete, and spray kicked up in a terrific display. Of the 22 K-84s dropped, analysis of post-strike video concluded that 12 found had their mark, with another four rated as "probable" hits. Still, the departing air crews were astonished to see the bridge still standing. But it was by no means intact. An after-action bomb damage assessment (BDA) by a trailing RF-4C Phantom revealed that several spans on the east side of the river had taken a severe beating, the surface so blasted and disjointed it rendered rail and road traffic impossible. It was a lesson well learned. For the remainder of Operation Linebacker, planners in need of precision ordnance, especially against point targets like the Doumer, would come to rely much more heavily on LGBs than their optically guided cousins. Still, precision munitions were both costly and rare. Unguided MK-82s, for example, cost about $4,000 apiece, while laser-guided MK-84s came in at four times that amount. Meanwhile, the number of Air Force Pave Knife laser designators, already extremely limited, would be cut to just

five functioning pods by July. For these reasons, conventional iron bombs would continue to carry the overwhelming majority of the workload for all but the most challenging targets.[21]

Just to the north, the Yen Vien strike force would have no need of such weapons. For an area target like Hanoi's principal railyard, old fashioned MK-82 "dumb" bombs would do just fine. The attackers pressed through the enemy air defenses, following the previously laid chaff corridor like a protective highway. As before, communist AAA gunners threw up a wall of flak. But the SAMs were strangely quiet. During the firing frenzy against the Doumer group, the crews had expended all loaded missiles in the bid to take down the Americans. But the Yen Vien group had followed on so quickly that the SAM teams had not had enough time to reload. The group's four SAM suppressing Wild Weasels of the 561st TFS did not see a single SA-2 airborne that day. To the south and east, eight 432nd TFW Phantoms established a barrier combat patrol (BARCAP) to shield against the MiG threat. Now nearing the target, the 16 F-4s of Ubon's 8th TFW pulled up in unison to 20,000 feet. Then, one after another, each flight of four dipped and dove, pickling the rail yard with 500-pound high explosive bombs. Dirt, steel, and rock came up in fiery blooms as more than 140 MK-82s did their work, chewing through the railyard's tracks, boxcars, and warehouses, leaving in their place a smoking ruin. Post-strike analysis later revealed that all through-tracks had been cut. As the Phantoms pulled up and away, the ground gunners chased them with dark blots of bursting AAA shells in a desperate bid for retribution. All 16 F-4s got away unscathed.[22]

The same could not be said for one of the group's escort fighters. As the Doumer and Yen Vien strike forces streaked west for the Laotian border and the waiting KC-135 refueling tankers, four flights of F-4s guarded the egress from various positions around the capital. For some minutes, all had remained quiet, with no reported bandit contacts. Harlow Flight, consisting of four F-4Es from the 336 TFS out of Udorn, was screening northwest of Hanoi. At about 1015, the flight passed near the airbase at Yen Bai. Hoping to mix it up, Harlow dropped to 8,000 feet to see if they could stir some interest. Suddenly, and with no radar warning, a J-6 appeared about 500 yards behind the flight's number two element. "Harlow 3, Harlow 3, you got a MiG on your ass!" came the frantic radio call. But it was a mistake. Even as the crew of Harlow 3 desperately scanned the airspace behind them, the J-6 pilot let loose a stream of 30 mm rounds into Harlow 4. Captains Jeffrey L. Harris and Dennis E. Wilkinson never knew what hit them. The heavy rounds sawed off the Phantom's left wing like a tender sapling, instantly sending the F- 4 into a steep dive. Moments later the aircraft simply piled nose first into the ground, exploding in a violent fireball. No one saw a chute; no one picked up a rescue beeper. Horrified, Harlow's flight commander torqued his Phantom into a violent turn in search of payback, blasting off a heat-seeking AIM-9E Sidewinder and two Sparrows in quick succession. All three missed. Then, the J-6 simply dipped away and was gone as suddenly as it had appeared. While the Sidewinder had proved itself an overall effective weapon system, its 9E variant—the first version designed specifically for the Air Force—was downright poor. Of the 71 air-to-air missiles fired during the campaign, only six 9Es would down their targets. As for Harris and Wilkinson, neither was seen alive again. In August 1978, the Socialist Republic of Vietnam (SRV) returned Captain Wilkinson's remains to U.S. custody. After three joint SRV-U.S. investigations and excavations between 1993 and 1996, the remains of Captain Harris came home as well. Still, both Air Force Linebacker strikes had been successful—while shooting down three MiGs in the process. But to the men who flew those missions, it was perhaps cold comfort in the face of losing three of their brothers.[23]

§

As morning turned to afternoon, the Navy's Task Force 77 kept up the pressure. At about 1219, *Constellation* launched its second Alpha strike of the day, this time against several strategic assets in Hai Duong, a town about midway between Hanoi and Haiphong. Hai Duong was home to a large POL facility, a major railyard, and a critical rail and highway bridge crossing. Cutting the line here would cripple transport between Haiphong and Hanoi to the west. *Connie's* 32-plane strike force consisted of the usual complement of EW radar jammers, combat air patrol and escort Phantoms, flak suppressing Iron Hands, and A-6 Intruder and A-7 Corsair attack bombers. As before, USS *Coral Sea* and *Kitty Hawk* launched similarly comprised follow-on groups at 10-minute intervals. By 1254, the *Constellation* strike group went feet dry near the mouth of the Red River and closed on the target from the south. As the flak suppressors worked over SA-2 and antiaircraft emplacements, the Intruders and Corsairs rained fire and steel on the railyard, 500-pound MK-82s kicking up a ferocious pyrotechnic display.[24]

On flak suppression that day were F-4 pilot Lt. Randall "Duke" Cunningham and his RIO, Lt. (junior grade) William "Willie Irish" Driscoll. Both were eager for a MiG fight—especially Cunningham, who only hours earlier had received a "Dear John" letter from his wife asking for a divorce. The airmen would soon get their wish. Ironically, Cunningham had not even been scheduled to fly the Hai Duong mission. But he was added at the last minute by CVA-9 Air Wing Commander Gus Eggert, who hoped to boost his moping lieutenant's spirits. Eggert even entrusted his personal Phantom, "Showtime 100," to the young airman. Now over target but unable to locate any AAA sites, Cunningham and wingman Lt. Brian Grant decided to unload their cumbersome Rockeyes on a few nearby warehouses. A lighter load, after all, meant a nimbler dogfighter. Just then, Red Crown radioed a frantic warning of a swarm of likely enemy radar contacts burning inbound. Driscoll glanced over his shoulder and spotted a mass of black dots on the horizon. It seemed the North Vietnamese were spoiling for a fight, as well. Within minutes, nearly two dozen enemy fighters had joined the fray. A glut of mostly MiG-17s—an antiquated, gun-equipped subsonic aircraft that was nevertheless a nimble and superb dogfighter—augmented by a few 19s and 21s, swarmed the Americans from every direction.[25]

Soon, the largest and most ferocious air-to-air battle of the war was underway, as some 45 U.S. and North Vietnamese fighters clashed in the skies east of Hanoi. Planes on both sides were soon blasted from the sky. The first to fall were Cmdr. Harry Blackburn and Lt. Stephen Rudloff from the *Connie's* VF-92. While the North Vietnamese would credit MiG-21 pilot Lt. Le Thanh Dao with the kill, the Navy maintains that Blackburn and Rudloff were downed by AAA fire. At any rate, the men successfully ejected, and their parachutes were observed by fellow pilots. A second *Constellation* F-4J, crewed by pilot Lt. Rod Dilworth and Lt. (junior grade) Jerry Hill, was severely damaged by what the Navy, again, determined was antiaircraft fire. In a remarkable feat, Dilworth, his right engine blown and gushing fuel, managed to nurse his crippled craft back to the *Connie*, where he affected a successful single-engine landing. The Phantom was a total loss, but Dilworth and Hill walked away unscathed.[26]

Cunningham and Driscoll, who had witnessed Blackburn and Rudloff go down in the midst of the chaotic fight, looked to even the score. In rapid succession, and with MiGs swarming all around, the duo first bagged a MiG-17 that was coming head on, guns blazing. As the MiG screeched past, Cunningham torqued hard into him, turned the tables, and let fly with an AIM-9G Sidewinder at a 1,000 feet. The MiG exploded in a terrific display. Cunningham then maneuvered to get position on a group of MiGs circling nearby when VF-96's executive officer Cmdr. Dwight Timm and his RIO Lt. James Fox went screeching by, two MiGs hot on their tail and another shadowing below. Angling into firing position, Cunningham called: "XO, reverse starboard!" If the executive officer did not break hard right, Cunningham worried his heat-seeking missiles would bag him and not the trailing MiG. Timm, unaware of the enemy fighter lurking below, instead leaned into his port turn believing he could shake the trailing MiG. "Showtime 112, reverse starboard! Goddamnit, reverse starboard!" This time Timm obliged, and Cunningham sent a Sidewinder up the MiG's tailpipe for his second kill of the day.[27]

Cunningham soon got word that the strike bombers were safely out of harm's way. The chaotic battle now winding down, the escort Phantoms began to peel away in ones and twos, trying to regroup for the push home. Themselves low on fuel and missiles, Cunningham and Driscoll banked east for the relative safety of the coast. But the North Vietnamese had no intention of making things easy. Cunningham spotted another MiG-17 coming head on, its 23 and 37 mm cannons blasting a steady fiery stream. Both fighters soon became embroiled in a twisting, turning dogfight, neither able to gain position on the other. But then the MiG pilot, himself likely low on fuel or ammunition, dipped hard and headed for the deck, trying to break contact. It would prove fatal. Now in better position, Cunningham let loose with another Sidewinder and watched the enemy plane explode in a bright orange fireball and plummet to earth. With their third kill of the day, Cunningham and Driscoll, who each owned two kills from previous, separate missions, became the first Americans of the Vietnam War to achieve coveted Fighter Ace status. The duo would remain the Navy's only fighter aces of the war.[28]

But they still had work to do. More MiG-17 and 21s swarmed as the two fought their way to the coast, fuel status now desperate. As they neared the beach, an SA-2 exploded about 500 feet overhead, peppering the Phantom with enough shrapnel to cripple its hydraulic system. Controls failing, Driscoll announced that Showtime 100 was fire, as well. Determined to get as far away from the hostile North Vietnamese coastline as possible, Cunningham somehow muscled the aircraft another 20 miles offshore before the men were forced to punch out. Eggert ordered five A-7s and another Phantom off the deck and instructed them to put in a patrol off the coastline. "Hit anything that comes off the beach," he radioed. After just 15 minutes of bobbing in the Gulf of Tonkin, the pair were picked up by a Marine rescue chopper off the USS *Okinawa*. News of their achievement rocketed through the *Connie's* passageways. By the time their Sea King rescue helo settled onto the carrier's deck a few hours later, a cheering crowd had gathered to welcome them home.[29]

Cunningham would be awarded the Navy Cross for his actions over Hai Duong, along with Silver Stars for separate 1972 engagements on 19 January and 8 May. Following his return from Vietnam in 1972, Cunningham went on to become an instructor at the Naval Air Station Miramar (TOPGUN). He retired with the rank of commander in 1987. As for Driscoll, he too would be awarded the Navy Cross for Hai Duong, as well as two Silver Stars and a Purple Heart for separate actions. Like Cunningham, Driscoll would become an instructor at TOPGUN, eventually retiring a commander.[30]

Cunningham and Driscoll were not the only Navy fliers to score kills in the fight over Hai Duong. Lieutenants Michael Connelly and Thomas Blonski of the *Constellation's* VF-96 were credited with two MiGs, with another going to Lt. Steven Shoemaker and Lt. (junior grade) Keith Crenshaw. *Coral Sea's* VF-51 scored another when Lts.

Kenneth Cannon and Roy Morris bagged their own MiG-17.[31]

And Task Force 77 still had more in store. At 1345, *Constellation* launched her third and final strike of the day, this time against the port facilities at Hon Gai some 20 miles northeast of Haiphong. At the same time, *Coral Sea* initiated her own final strike, targeting the rail and highway bridge at the port of Cam Pha just up the coast. A short while later, *Kitty Hawk* joined the fray when it launched an armed reconnaissance over the coastal area near Cam Pha, destroying several storage facilities in the port city. Perhaps due to the heavy pressure exerted by earlier strikes, very little resistance was encountered. Aside from two MiGs that made a brief appearance, the strike groups enjoyed a mostly free hand as they carpeted targets with 500-pound MK-82s. One escort Phantom, crewed by Lt. Curt Dose and Lt. Cmdr. Jim McDevitt of *Connie's* VF-92, did have a brutally close call when two SAMs rocketed to within killing distance of their F-4. Neither exploded, however, and simply flew on past. Meanwhile, in the harbor at Hon Gai, a potentially catastrophic incident occurred when two errant MK-82s struck and severely damaged the Soviet cargo ship *Grisha Akopyan*, killing one crewman and injuring its captain. Soviet officials complained, but nothing became of the incident. By 1700, the last of *Kitty Hawk's* planes were headed home. And so ended Operation Linebacker's first frenetic day of action.[32]

U.S. planes flew 414 attack and support sorties on 10 May, including 294 for the Navy and 120 for the Air Force. Several factors explain the disparity in sorties flown. First, the Air Force's area of responsibility included Hanoi, one of the world's most heavily defended cities against air attack. The capital, for example, was home to 10 missile battalions of the 361st Air Defense Division, while just four defended Haiphong. Likewise, all four of North Vietnam's top-line fighter regiments were concentrated at airfields just northwest and northeast of the capital. Next, with the exception of Bien Hoa Airbase just north of Saigon, all USAF air assets had by May 1972 fully transferred to various facilities throughout Thailand. This substantially increased the distance between USAF bases and targets in North Vietnam. Finally, the Air Force was primarily responsible for interdiction and close air support missions in South Vietnam, as well as the border areas of Laos and Cambodia. These included tens of thousands of B-52 sorties throughout Southeast Asia. The combined effort would constitute the bulk of the Air Force's workload throughout Linebacker. All of these factors conspired to further complicate mission design and resource allocation, requiring Air Force planners to assemble and allocate a higher proportion of escort fighters, antiaircraft suppression assets, refueling tankers, and so on for each mission.[33]

Target selection was based on three broad objectives: destroy the North's transportation, logistics, and supply networks, wreck key segments of its military-industrial base, and degrade its air-defense capabilities. All, it was believed, would imperil Hanoi's ongoing invasion of the South while convincing Northern leadership to resume serious peace negotiations. To that end, Linebacker seemed off to a good start. Opening-round strikes by the Air Force and Navy knocked out of commission four major bridges between Hanoi and Haiphong, crippling rail traffic between the all-important cities. Meanwhile, logistics hubs, POL facilities, railyards, and other storage and transportation assets all suffered varying degrees of destruction. In the process, strike mission escorts, air-defense suppression aircraft, and combat air patrols had already taken a heavy toll on North Vietnam's air defense network. All of these efforts would only intensify as Operation Linebacker unfolded in coming months.[34]

As expected, the North Vietnamese threw up a spirited defense, launching nearly 100 SA-2s, firing untold AAA and smaller-caliber rounds, and sending up some 41 MiGs to challenge U.S. pilots. The latter figure was probably an underestimate since it likely represented only those enemy aircraft actually seen by American fliers. On what was the war's most intense day of air-to-air combat, Navy pilots shot down eight MiGs, including seven in the fight over Hai Duong. In return, the North Vietnamese downed two Navy Phantoms, with a third so badly damaged that it was ultimately scrapped. According to the Navy, all fell victim to antiaircraft fire. Lt. Steven Rudloff, the VF-92 RIO shot down during the fight over Hai Duong, survived his ejection only to be captured a short time later and taken to Hanoi's notorious Hoa Lo prison.[35]

The place was a holdover from the French, who had built and used it as a colonial prison since the 1880s. Called the *Maison Centrale*—the Central House—it had served as an infamous site of brutality and persecution for Vietnamese political prisoners dubbed most dangerous by colonial authorities. But the communists had beaten the French at Dien Bien Phu in May 1954, and had since made Hoa Lo the center of their own prison system. The Vietnamese called it "hell's hole," perhaps owing to the large number of fiery cook stove shops that traditionally lined Hoa Lo Street. But since August 1964 and the capture of Navy flier Lt. (junior grade) Everett Alvarez—the first American pilot shot down over North Vietnam— the original translation took on a very different meaning. For the North Vietnamese had revived it as a center of hellish interrogation, torture, and privation—this time for captured U.S. servicemen. Apropos of the gallows humor often found among men living in extremis, the inmates had sardonically dubbed it the "Hanoi Hilton" after the posh hotel chain back in the States. It was, of course, anything but. By war's end, some 600 Americans would suffer nearly unimaginable mental, physical, and spiritual cruelty as "guests" of the Hilton.[36]

What happened to Rudloff's pilot, Cmdr. Harry Blackburn, is less clear. All that is known is that he perished under mysterious circumstances sometime after ejecting. The Vietnamese government returned his remains to U.S. custody in 1986. For its part, the Air Force scored three MiG kills that day, but two of its Phantoms were downed by gun-equipped J-6s, killing three crewmembers and consigning the fourth to a harrowing 23-day fight for survival. The Triple Nickel's Capt. Roger Locher would endure his ordeal and be rescued on 2 June. He was one of the lucky ones. As Operation Linebacker intensified in coming weeks and months, more and more American airmen would put their lives on the line in the skies over North Vietnam.[37]

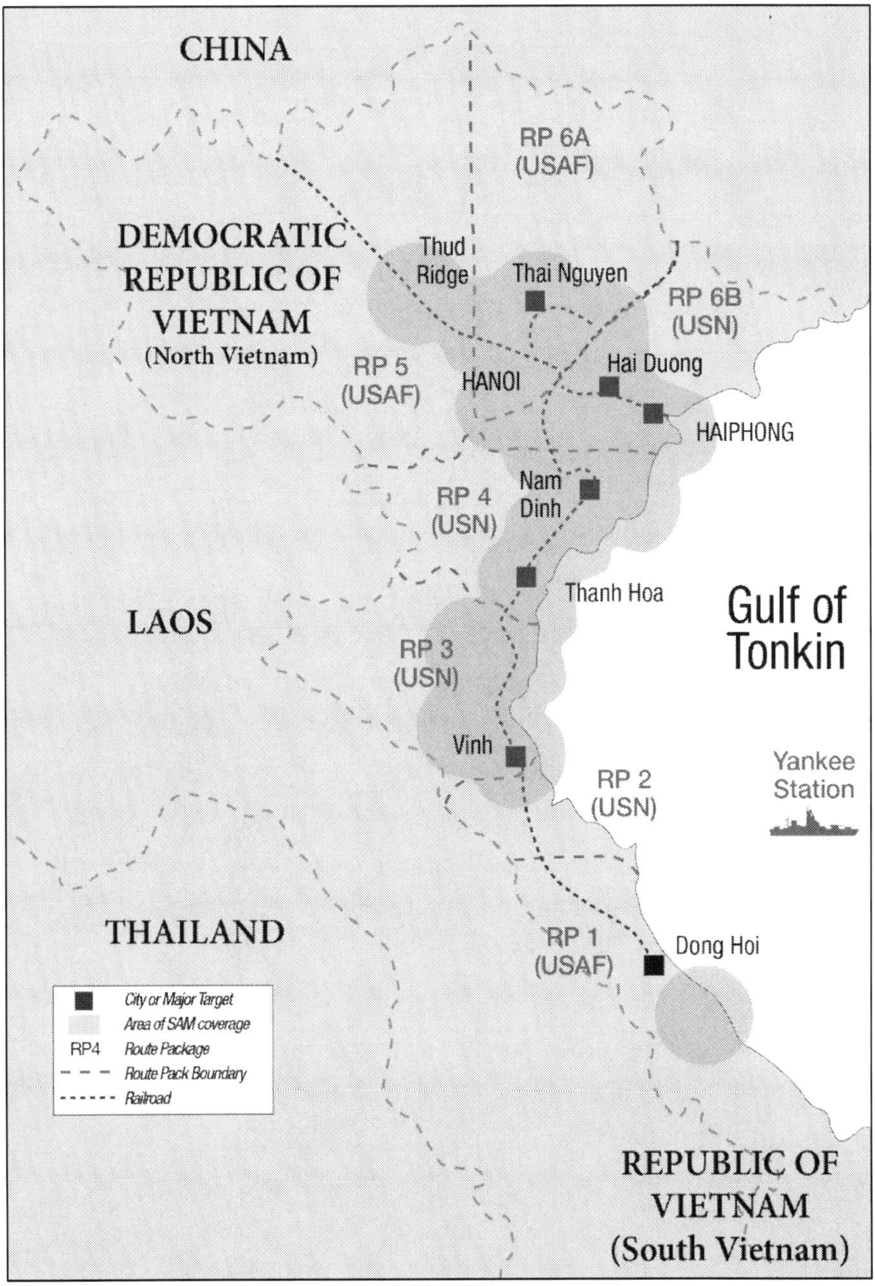

Figure 3: Route Packages

5

STEEL RAIN

11 May – July

Despite the frenetic action of Linebacker's first full day, the campaign's tempo and ferocity would only intensify in coming weeks and months. On 11 May, and before the North Vietnamese could affect repairs, USAF Gen. Vogt dispatched a strike force to finish off Hanoi's Paul Doumer Bridge, badly damaged but still standing following the previous day's action. In yet another indicator of the value of precision-guided weaponry, the follow-on Doumer group included a strike element of just four LGB- equipped Phantoms. Such a mission four years earlier would have required 10 times that number. The strike Phantoms' 2,000 and 3,000-pounders again scored hits, this time blasting three of the Doumer's spans from their abutments and knocking it out of action. As the North Vietnamese feverishly worked to affect repairs in coming months, the Air Force would return repeatedly to ensure the Doumer stayed out of commission. On 13 May, the Air Force reappeared to dispose of another vital transportation link, this time the Thanh Hoa Bridge—the notorious Dragon's Jaw. Located about 106 miles south of Hanoi, the massive structure had eluded destruction by U.S. pilots since 1965. Traversing the Song Ma River and supporting the only rail line in North Vietnam's panhandle, the Dragon's Jaw had for years been indispensable in transporting immense quantities of materiel from the Hanoi-Haiphong complex to North Vietnamese forces in the South. It had most recently been hit by the 8th TFW on 27 April as part of Operation Freedom Train. As with the Doumer, the strike's 2,000-pound EOGBs had temporarily closed the bridge to traffic but had left it largely intact. This time, 14 F-4s of the 8th TFW came armed with 2,000 and 3,000-pound LGBs, along with standard free-fall MK-82s. In all, the Air Force released nearly 70 tons of high explosives, breaking the Dragon's Jaw and knocking it out for the long haul. The bridge would meet the same fate as the Doumer, as periodic strike missions returned to ensure it remained out of action for the remainder of the war.[1]

The hits kept coming throughout May as strike authorizations widened to include industrial and infrastructural targets. Planners placed special emphasis on cutting the flow of fuel to Hanoi's mechanized forces in the South. In mid-May, the USS *Saratoga* (CV-60) arrived on Yankee Station and was soon tasked with hitting a series of fuel routing junctions in North Vietnam's southern panhandle. The area boasted only light air defenses, but aircrews soon discovered that the region presented its own unique challenges. Lt. Craig Honour, an F-4J driver with the VF-31 Tomcatters, recalls his Combat Information Center (CIC) briefing. "They've got all these beautiful pictures of some manifold complex, about 10 or 15 different pipes coming in from different directions," he says. "They all connect and valves redistribute the fuel and send it every which way." There was one problem, however: the reconnaissance pictures were nearly two years old. "Everybody just burst out laughing," Honour says. "We're like, 'Are you really expecting us to be able to find this thing?' That's when I realized the strategic thinkers in the command authority didn't have a clue."[2]

Sure enough, when Honour's four-ship element arrived on station, there was nothing to see but a thick blanket of jungle stretching to the horizon. Here and there, limestone karsts jutted like stone fingers through the vegetation. The aircrews settled on improvisation, using the rock formations to roughly triangulate where they were supposed to hit. "We had an accuracy expectation of about two miles…for a target roughly 50 feet across," says Honour. "It was like dropping a dart off the top of the Empire State Building and hoping to hit a dartboard on the sidewalk." Nevertheless, the Tomcatters dutifully dropped their ordnance and reported back: the forest canopy had taken quite a beating, but there were no secondary explosions to indicate that anything other than trees had been hit. Still, five days into the fuel interdiction campaign, MACV announced that U.S. strikes had at least managed to take out all major North Vietnamese pumping stations in the southern panhandle.[3]

To further degrade the MiG threat, Air Force F-4s on 16 May struck Hanoi's main air defense command and control complex at Bach Mai Airfield, "wrecking" its headquarters and destroying several buildings.[4] Meanwhile, naval strikes on crucial rail and road bridges along Highway 1 from Vinh to the DMZ helped curtail the flow of men and materiel for the North's Nguyen Hue Offensive in the South. Further strikes managed to knock out another four rail bridges linking Hanoi with Haiphong that month. On 23 May, Air Force and Navy pilots flew 190 strikes against the North Vietnamese heartland. Chief among them were Navy attacks on the power plants at Hongai, about 24 miles northeast of Haiphong, and Nam Dinh some 90 miles to the southwest. Meanwhile, a flight of four F-4E gun-equipped Phantoms of the 35th TFS were flying MIGCAP near Kep Airfield when they got in a scrap with up to eight MiGs. Air Force pilots bagged four enemy fighters in the exchange.[5]

The Navy was getting its share of MiG kills, too. On 18 May, two F-4Bs with the USS *Midway*'s VF-161 Chargers were flying MIGCAP for an Alpha strike on Haiphong's railroad bridge when they downed two MiG-19s with heatseeking Sidewinders. The Chargers were led by Cmdr. Wayne "Deacon" Connell, an officer who insisted his squadron imbibe the warrior ethos. The Chargers were to be MiG-killing machines, first and foremost. "The skipper wanted a MiG even worse than I did," recalled Chargers pilot Lt. Pete La Chat. "His attitude was, 'A MiG on your six is better than no MiG at all.'" Although the F-4 was originally designed as an over- the-horizon interceptor—more suited to shooting down Soviet nuclear-armed bombers than tangling with MiGs—Connell knew the realities of Vietnam required his men to become expert dogfighters, too.[6]

The skipper charged Lt. Cmdr. Ronald "Mugs" McKeown with ensuring aircrews were up to snuff. McKeown was a supremely confident jet jock who only half-jokingly handed out business cards identifying himself as "the world's greatest fighter pilot." He may not have been far off. McKeown had been a test pilot for the Navy's VX-4, Air Test and Evaluation Squadron at NAS Point Magu, CA, where he helped develop air tactics and battle maneuvers for the Phantom. He knew his way around enemy aircraft, too. As part of the Defense Intelligence Agency's (DIA) Have Doughnut/Have Drill projects in the late 1960s, McKeown had flown MiG-17s and 21s, which Israel had acquired by various means, in simulated dogfights against U.S. fighters over Groom Lake, NV. There he learned the strengths and weaknesses of enemy aircraft—and the best ways to beat them. McKeown knew there was no way the F-4 could out turn the nimbler Soviet fighter horizontally. "Guys died trying," said La Chat. Instead, the trick was to go vertical. McKeown advised his pilots to use the Phantom's more powerful thrust to outclimb the enemy. The so-called "yo-yo" maneuver epitomized the concept: Roar up above and behind a MiG, attain a favorable firing solution, then dive for the kill. McKeown and his longtime radar intercept officer (RIO) Lt. Jack "Fingers" Ensch, would soon put those dogfighting skills to the test.[7]

On 23 May, McKeown and Ensch, callsign Rock River 100 and their Rock River 112 wingman, had just gone feet dry on their way to a MIGCAP station north of Haiphong when their headsets crackled. 'We have bandits, 2-7-0, 30 miles and closing.' It was Red Crown, the PIRAZ ship in the Gulf. The airmen's ears pricked up. The radarman had said "bandit" not "bogey." The latter was merely an unidentified contact and could even be a friendly aircraft. 'Say again, Red Crown. Did you say bandits?' McKeown called. 'Affirmative. You are cleared to arm and fire.' Rules of engagement usually required pilots to visually identify a target before firing. The aircraft burning inbound had been positively identified as enemy fighters. The RIO's adrenaline surged. They had conducted too many dogfighting simulations to count. Practice, after all, helped develop a kind of muscle memory. "It's like baseball," says Ensch. "Simulated fights are like batting practice. But now we were getting ready to hit for real in a big league game."[8]

Each Phantom carried four radar-guided Sparrows and four heat-seeking Sidewinders. Neither, however, was gun-equipped. No matter. Now that they knew the contacts were hostile, the Sparrow would do the trick before they ever saw the MiGs. Ensch kept his eyes glued to the radar screen. The RIO picked up a few paints on one of the MiGs, but the ground clutter was preventing a solid lock. Rock River 112 radioed over. 'My radar just went tits up. I'm blind.' That would make things that much harder. Suddenly, McKeown called out, 'Tally-ho on the nose!' He had spotted the bandits straight ahead, five miles and closing fast.[9]

But before McKeown could react, the two MiG-19s had already screeched past, splitting between the Phantoms. 'Cross trails,' radioed McKeown. 'You go low, we'll go high.' The Phantoms turned in on each other, simultaneously clearing one another's six o'clock while pivoting to pursue the enemy fighters. "We're thinking, 'Hot damn! Two of our boys just shot down a couple MiG-19s last week,'" says Ensch. "We're like, 'We're gonna get us a couple!'" But the 19s were not alone. Neither Red Crown nor Ensch had picked up a flight of four MiG-17s trailing close behind. The Chargers had turned right into the midst of the enemy formation. 'It's raining MiGs!' called McKeown. The blunder might have been fatal had the North Vietnamese not made a critical error of their own. The trailing fighters had not kept enough distance between themselves and the MiG-19s. If they had, McKeown and Rock River 112 would have never known what hit them. Instead, the Chargers were now intermingled with the MiG-17s, outnumbered three to one, and far from out of the woods. "That's when the shit hit the fan," Ensch says.[10]

Two of the 17s followed Rock River 112 down low, while the other two went for McKeown and Ensch. Meanwhile, the 19s were turning back into the fray. They were outnumbered four to one. Too close for Sparrows now, the missiles were nothing more than dead weight. This would be a Sidewinder fight. McKeown maneuvered to gain a firing solution, while Ensch kept his head on a swivel. He spied a 17 moving into shooting position on their eight o'clock. "Mugs! This guy's getting ready to shoot!" Ensch called. Without a moment's hesitation, McKeown jammed the rudder in and hauled back on the stick, throwing the Phantom into a wild backflip. In aviator parlance, the aircraft had "departed"—gone completely out of control. But it was also completely by design. McKeown called it his "last ditch" move, and it was a maneuver he had practiced over and over with VX-4. But he had never done it with Ensch in the backseat. "It was like being inside a Tilt-a-Whirl, like a ride at Disney Land," Ensch says, laughing. "Needless to say, the MiG didn't try to follow us on that maneuver."[11]

And now hunter had become prey. McKeown's improbable

move put the Phantom right on the MiG's six o'clock. He let loose with a Sidewinder. But the MiG pilot had a few moves of his own, jinking wildly to evade the missile. McKeown maneuvered behind a second 17 and fired, but it too defeated the Sidewinder. Now the 19s, notorious for their limited fuel capacity, were fleeing the area. But it was still a two-on-one fight, and another MiG-17 had come up hard on McKeown's tail, cannons blazing. "Mugs! Four o'clock, tracking and shooting," hollered Ensch. "This guy's really gotten in on us!" Ensch watched the 37 mm rounds, fat fiery golf balls arcing closer to the Phantom. McKeown immediately turned inside the pursuing MiG to break its firing solution. The enemy pilot pulled his nose up, no doubt confident he could stay inside the Phantom's turning radius and regain its six. It was a grave mistake.[12]

Now the hours spent in the MiG-17's cockpit served McKeown well. He knew that in that position the jet's bulbous nose would block the pilot's view. For a few precious moments, the Phantom had become "invisible." It was just the opportunity he needed. McKeown cut the throttle and went full forward stick, pulling negative Gs and gaining separation as the 17 roared past. McKeown then came back hard on the throttle, sweeping upon the MiG's six o'clock. The Sidewinder's seeking tone went from a soft warble to an urgent growl in his headset—solid lock. The AIM-9 tracked right into the 17's tail, blasting the aircraft apart and sending the MiG pilot ejecting for his life.[13]

Meanwhile, Rock River 112 had spent the entire engagement maneuvering wildly to keep the other two 17s off their tail. Most of the action had taken place at tree-top level—the worst possible place for a Phantom to fight the much-nimbler MiG. McKeown keyed his mic: 'Come east, hold and cross,' he radioed. That heading would give him the chance to hit the MiG from above and behind. It worked. Rock River 112 led the trailing 17 into perfect position. McKeown dove and let loose with his last Sidewinder, taking the MiG in the tail and saving Rock River 112 in the process. The remaining MiGs had seen enough and scattered toward the horizon. Ensch suddenly became aware of chatter in his headset. It was Red Crown. 'Get out of there, get out of there!' came the call. The Alpha strike was long over, and the attack element had already gone feet wet over the Gulf. There was no telling how long the PIRAZ had been calling. "I don't remember any of it," Ensch says. "We were so busy with the fight, I didn't hear a thing."[14]

Only when McKeown and Ensch had cleared the beach did reality finally hit. "We're pounding the sides of the canopy and screaming, 'God damn! Do you realize what we just did? We fought a dog fight and shot down two MiGs for Chrissakes!'" McKeown and Ensch were the last aircraft to recover that day. Knowing that word of their double-MiG kill had already spread, McKeown decided to give the *Midway* a fly-by to remember. The pilot screeched in lower and lower at some 450 miles per hour, buzzing the ship at flight-deck level. "And I'm in the backseat saying, 'Mugs, goddamnit! If you kill us here I'll never speak to you again,'" says Ensch, laughing. At the last moment, McKeown pulled the F-4 into a victory roll off the port side. The elated *Midway* crew roared its approval. McKeown and Ensch would be awarded the Navy Cross for their actions that day.[15]

In August, McKeown left to assume command of the Navy's TOPGUN fighter weapons school at Naval Air Station Miramar. Just 13 days later, Ensch would be shot down by an SA-2 and captured on 25 August. His pilot that day, Lt. Cmdr. Mike Doyle, was killed. Throughout Ensch's 216 days of captivity, McKeown refused to wear his Navy Cross. He and Ensch had earned the honor together, and they would wear it together. Ensch was eventually released from the Hanoi Hilton on 29 March 1973. In January 1974, the RIO reunited with McKeown as an instructor and executive officer with TOPGUN. Both would attain the rank of captain and go on to long and illustrious careers. As for VF-161, the Chargers would score five kills over the final eight months, including the Vietnam War's last MiG shootdown on 12 January 1973.[16]

§

Operation Pocket Money was paying dividends, as well. On 16 May, a flotilla of eight Soviet warships approached to within 200 miles of the North Vietnamese coast as 13 of its cargo ships were enroute to Haiphong. But signaling that Moscow had no interest in pushing the issue, the Soviet warships melted away two days later, the cargo ships diverting to "neutral" ports. Unknown to U.S. officials at the time, Soviet leadership had since the earliest days of the war expected the U.S. to close North Vietnamese ports. According to then-Soviet Ambassador to the United States Anatoly Dobrynin's 1995 memoirs, Moscow had no interest in seriously challenging any U.S. blockade, fearing it would unnecessarily escalate tensions between the superpowers to the benefit of their rivals in Peking.[17]

With some 85 percent of North Vietnam's 150,000 tons of monthly imports arriving by sea, the port closures had forced resupply to divert almost exclusively to just two rail lines and eight major roadways leading from China. Soviet and other Eastern Bloc nations were now forced to deposit virtually all goods bound for North Vietnam—including the overwhelming majority of its petroleum and sophisticated weaponry—at PRC ports. Despite Soviet-Sino tensions, the Chinese agreed to assume responsibility for rail and road transport into North Vietnam. Again unknown to U.S. officials, this contingency plan had been in place since at least 1967. According to a memo of conversation released by the PRC in the late 1990s, then-Chinese

Premier Zhou Enlai informed Hanoi that, should the U.S. close sea imports, China was not interested in directly intervening in the war. Instead, the PRC pledged to ensure supplies continued to flow through China.[18]

Nevertheless, the vastly constrained transport system presented a target-rich environment for U.S. airpower. Responsibility for interdicting overland supplies fell chiefly to the 8th TFW out of Ubon. As May unfolded, the Air Force opened strikes on 13 rail and highway bridges connecting North Vietnam's northeast and northwest rail corridors with China. Many of these targets spanned deep Annamite Mountain gorges and could not easily or quickly be repaired, especially under relentless air attack. "As fast as they would build them, we would knock them out again," Gen. Vogt later said. An especially juicy target was the Lan Giai rail bridge in the northeast. Lan Giai had long been off limits because it was well within the U.S.'s self-imposed 30-mile buffer along the Chinese border. On 25 May, after receiving special permission from the Joint Chiefs of Staff (JCS), an 8th TFW strike group hit the bridge with LGBs and EOGBs. Heavy cloud cover again foiled the optically guided munitions, but 19 of the 30 LGBs scored direct hits, knocking down six of the bridge's 11 spans. As more rail lines were severed, hundreds of supply-laden boxcars languished cut off and immobile. According to U.S. analyses, by early June some 600 boxcars were left stranded throughout the country. The bombing of border crossings would continue unabated until the PRC on 12 June issued its strongest condemnation since the start of Linebacker, calling the near-border strikes "grave provocations against the Chinese people." Two weeks later, the Nixon administration relented and reestablished a 25-mile buffer zone along the Chinese border.[19]

Navy surface vessels were getting into the act, as well. Closing the harbors and ports had added another level of complexity and danger to the North's import lifeline. Closing Haiphong alone had cut total imports by 30 percent. Rather than using port facilities to efficiently offload the average 6,000 tons of cargo for each foreign vessel, North Vietnam was now forced to unload offshore. This was accomplished by manually offloading cargo onto small, shallow draft boats and barges and then ferrying the badly needed supplies to the beach. Aside from vastly increasing the time, labor, and resources required, this new reality also made the boats vulnerable to air and sea attack. And so it was that on 17 May, the light cruiser USS *Oklahoma City* happened upon a flotilla of North Vietnamese supply boats as it neared the mouth of the Cua Viet River. Likely trying to resupply NVA troops that had recently seized South Vietnam's northernmost Quang Tri Province during the Easter Offensive, the boats were a ripe target. The *Oklahoma City* let loose with her six- inch guns, sinking 10 of the supply vessels, while damaging 20 others. The trend would continue. By mid-June, the naval blockade, along with opportunistic air attacks, would destroy some 1,000 North Vietnamese supply and other surface vessels.[20]

As May came to a close, Linebacker pilots had flown more than 6,000 sorties, hitting rail lines, bridges, pipelines, power plants, military training facilities, MiG airbases, antiaircraft emplacements, and command and control air assets throughout North Vietnam. U.S. aircrews had downed some 27 North Vietnamese MiGs but had lost 20 of their own aircraft to ground fire and enemy fighters. In all, 11 Navy and 9 Air Force planes went down, with nine aircrew members killed and another 12 captured. Still, Linebacker showed no signs of letting up.[21]

By early June, U.S. air assets continued to pour into Southeast Asia. The number of Air Force fighter-bombers had tripled since the start of the year, and the Navy had gone from two aircraft carriers and their attendant air wings to six, with another two carriers on the way.

Meanwhile, the complement of Air Force B-52s in theater had quadrupled.[22] On 7 and 8 June, Air Force planners, citing markedly reduced SA-2 launches in previous weeks, ordered the first B-52 strikes over North Vietnam since Operation Freedom Train seven weeks earlier. Operating in three-ship cells, the "big ugly fat fellows" (BUFFs), as they were affectionately known by the "crew dogs" who manned them, dumped 30 tons of high explosives apiece on NVA troops concentrations just north of the DMZ, rumbling the earth for miles in every direction. Meanwhile, fighter-bombers using EOGBs and LGBs collapsed the Luong Truong rail tunnel near the Chinese border about 95 miles northeast of Hanoi. Unlike the mass- firings during the campaign's first few days, not a single SAM launch was recorded. By 4 June, U.S. pilots were reporting fewer than 10 launches per day throughout the country, indicating the North was likely conserving stocks with so little hope of resupply.[23]

On 10 June, the Air Force carried out a highly orchestrated attack on Hanoi's newest and largest hydroelectrical plant at Lang Chi. Located about 65 miles northwest of Hanoi on the Red River, the facility was connected to a dam that used water turbines to create electricity. The Soviet-built, 122,000-watt facility was capable of supplying Hanoi with some 75 percent of the electricity required for its industrial and defense needs. This mitigated the impact wrought by targeting other power plants, so something had to be done. Washington knew it was vital to attack Lang Chi, but the strikes would have to be exquisitely precise. Tens of thousands of civilians lived near the facility, and analysts estimated that some 23,000 would die if the dam were breached and flooded the area. Encouraged by the efficacy of their new laser-guided munitions, Air Force planners believed the mission could be accomplished. Washington gave the go-ahead, and an 8th TFW strike group, including a dozen F-4s armed with 2,000-pound LGBs, hit the facility in two waves. The first used the precision bombs to gouge a hole in the plant's 50 x 100-foot concrete roof. The second wave then dropped 12 MK-84 fuse-delayed LGBs through the hole, destroying the facility's four turbines and generators without damaging the dam or spillway beneath. It was a remarkable feat.[24]

The air war only intensified in coming weeks. On 13 June, U.S. pilots flew some 340 sorties over North Vietnam—the most in the war's history—knocking out or severely damaging 10 bridges, including four northwest rail lines linking China with Hanoi. Also included was the rail and highway bridge at Hai Duong, severing the main rail link between the capital and Haiphong. Other strikes in northwest Hanoi knocked out the only factory producing pontoon bridges to replace those recently destroyed.[25] The Nixon administration ordered a four-day bombing halt in the Hanoi area in light of Soviet Head of State Nikolai Podgorny's visit. Fearing North Vietnam would use the respite to regroup and strengthen its air defenses—as had happened with similar pauses during Rolling Thunder—the move drew strong protests from air commanders. The air strikes resumed with gusto on 18 June as U.S. aircraft matched the previous record of 340 sorties flown over North Vietnam. Attacks included strikes on the MiG bases at Kep, Bai Thuong, and Quang Lang, cratering runways and temporarily knocking them out of action. Unfortunately for U.S. planners, however, attacks on MiG bases had proved more harassment than an expectation that bases would remain out of service indefinitely. North Vietnam had shown a remarkable ability to swiftly repair even severely damaged airfields. But the attacks kept the enemy off balance, requiring Hanoi to devote considerable resources and labor for repair.[26]

As June came to a close, strike elements from the USS *Saratoga* (CV-60) and its Air Wing 3 used Walleye EOGBs to blast the rail and highway bridge over the Song Day River at Phu Ly, 40 miles south of Hanoi. The group also wiped out some 75 percent of the area's military supplies by destroying nearby storage facilities with free-fall ordnance. The *Saratoga,* on its first and only deployment to Vietnam, would suffer considerable losses during her appearance, with 13 aircraft downed in combat, and another four lost to accidents. The Air Force rounded out the month by using LGBs to take out the steel plant at Thai Nguyen on 24 June. It was North Vietnam's only facility capable of producing the structural steel required to build and repair bridges, railroads, and other complex structures. The next day the Air Force knocked the lights out in Hanoi by bombing its Viet Tri power plant. On 27 June, U.S. aircraft flew 320 strike sorties over North Vietnam, most right into the teeth of ferocious antiaircraft fire over the Hanoi-Haiphong corridor. This included the previously off limits Bac Mai Airfield near the capital's downtown.[27]

523rd TFS squadron commander Lt. Col. Farrell "Sully" Sullivan, 37 and his WSO Capt. Richard "Dick" Francis, 29, out of Udorn had the unenviable task of spearheading one formation right into the heart Hanoi air defenses. Sullivan's four-ship chaff flight was responsible for dispersing clouds of aluminum strips to help confuse antiaircraft radar. Chaff worked by tricking enemy ground stations into "seeing" hundreds of false targets. "They can't tell the real ones from the fake ones…except for one," says Francis. "The guy at the front of the chaff cloud." Worse yet, chaff pilots were required to maintain slow, steady, level flight to ensure the metallic clouds dispersed properly. Any deviation—even evasive action to dodge murderous groundfire—would put the strike element at risk. Chaff Phantoms' only protection were the AN/ALQ-87 radar- jamming pods mounted under the fuselage—cold comfort in the face of withering AAA fire and SA-2 launches. "Sully had briefed us that we were going to rely on these jammers," recalls Francis. "He said, 'If we get a SAM coming up at us, don't worry about it. Hold your position; we're going to let the jammer do its job.' And we're thinking, 'Boy, I'm not liking this much.' But that was the way you had to do it. And that's exactly what we did."[28]

At about 0900, the chaff formation neared the center of Hanoi. Francis watched as several SA-2s detonated to his left and right—too far to cause damage, close enough to know ground crews were targeting them. "On a scale of one to 10, I'd say the 'pucker factor' was about a 15 at that moment," he says with a chuckle. Still, the jamming pods seemed to be doing the job. Sullivan's formation pressed on, slow and steady, clouds of aluminum strips spreading in its wake. At last, the flight neared its turn for home, and a wave of relief washed over Francis. He hit the Inertial Navigation System (INS) reset switch, giving Sullivan the egress heading. But just as the pilot began his right turn, Francis's headset came alive with an urgent warble, buzzing like an angry rattlesnake. RADAR LOCK. 'SAM ten o'clock! SAM ten o'clock low!' someone called. Francis routinely flew with his flight harness loosened so he could lift himself up to scan for threats. But just as the Wizzo pulled up to see over the left engine intake, the Phantom's canopy exploded in a tornadic whirlwind of acylic and metal. "It felt like we'd been hit by a Mack Truck," he says. "Rapid decompression, and suddenly we're tumbling out of control." The WSO called to his pilot over the wind's deafening roar. "Sully, can you hear me?" No response. "Sully!" Again, no reply. Francis tried the stick, desperate to right the ship. Nothing there. All hydraulics gone. "Then I felt fire all around me. I knew it was time to get out."[29]

In the F-4, a Command Selector Valve (CSV) above the rear instrument panel governed the order of aircrew ejection. As aircraft commander, Sullivan preferred to have full control over the ejection sequence, so he had directed Francis to set the CSV to "solo." That prevented the WSO from inadvertently ejecting his pilot, while still allowing the backseater to exit the craft on his own. Francis groped for the ejection handle between his knees. But the Phantom was tumbling wildly, positive and negative Gs throwing the WSO violently back and forth in his loosened harness. After several flailing attempts, Francis at last gained purchase and pulled with all his might. A split second later he blasted free of the stricken craft and into the bright blue skies 16,000 feet above Hanoi.[30]

The force of the ejection had driven all the blood into Francis's lower extremities. On the verge of passing out, he somehow remembered his training, tightening his muscles and breathing to force the blood back to his brain. As his vision cleared, the WSO realized with horror that his chute had not opened, and for a wild moment lost track of just how long he had been falling. But then the chute suddenly deployed. He looked left and right but could spot no sign of Sullivan's parachute. Just then, an F-4 on an attack run screamed past. 'Holy shit!' he thought. 'I'm right over the target. I'm gonna get run over up here.' The WSO needed to maneuver the chute and fast. He pulled at a set of dangling red lanyards, releasing a quartet of risers, which allowed some control over steering. Just then, a torrent of red tracers flicked past. "I'm thinking, 'Holy crap, they're going to shoot me in the chute,'" says Francis. Then he realized that the "tracers" were actually rivulets of blood streaming from overhead. He looked up and spied an ugly shrapnel wound in his right biceps. Another jagged shard had punched into his upper right back. The voice of a crusty survival training sergeant at Fairchild Air Force Base echoed from the past. *There's a one-in- three chance you will experience a combat ejection,* said the voice. *So you need to know where your survival equipment is because you may have to perform first aid on yourself while in the chute.* Francis groped at his flight harness until he found the first aid kit. He pulled a tourniquet and sinched it tight above the shrapnel wound.[31]

Below, the city rushed up to meet him. There was no way he wanted to come down in the middle of Hanoi. But the Tonkin Gulf was

80 miles away. 'At least get to the countryside,' he thought. A ripping sound from above dispelled that hope. An entire parachute panel had suddenly torn away. The chute riggers had told him that was next to impossible. So much for that. Now he was falling faster than ever. Francis managed to steer some five miles southeast of the city before his time ran out. He soon realized he would come down in a North Vietnamese truck park, an open-ended metal shed covering a few of the vehicles. And if Francis did not do something, he would smash straight into its rusty tin roof. With one final effort, the WSO pulled past the structure, instead crashing down near a series of concrete parking chocs. Before he could move, a wild-eyed NVA soldier was already shoving the muzzle of an AK-47 into his throat. A half dozen more soon arrived. The soldiers stripped him to his underwear, bound his arms behind his back, and tossed the Wizzo into the back of a flatbed truck. Above, the air war raged on, as an F- 4 rolled on a target. The NVA dove for cover in a nearby drainage ditch. "I'm there on my back thinking, 'Have I got a front row seat to the war!'" he says chuckling.[32]

Francis would be taken to a place its inmates called "The Zoo," a prison camp in southwestern Hanoi. It was so named because its cell bars allowed guards to peer inside like spectators at zoo animals. He would remain a captive of North Vietnam for 275 days until his release on 28 March 1973. Francis was awarded the Silver Star for his actions on 27 June 1972, along with a Bronze Star and Distinguished Flying Cross for other service during the war. The WSO went on to earn his pilot's wings in 1975, retiring a lieutenant colonel in March 1986.[33]

As for Sullivan, he would remain MIA until a June 1983 American-Vietnamese investigative team recovered human remains from a burial site north of Hanoi. Testing by the Central Identification Laboratory-Hawaii identified the remains as Sullivan's on 27 June 1983—11 years to the day of his shootdown. 1st Lt. Fred Hastings, a

WSO who was on Sullivan's flight that day, later recalled him as a man who led by example. "Rather than assign it to his underlings, he flew the most dangerous missions," Hastings wrote. "He never expected anyone to do anymore or anything he would not do."[34] MiG pilots and SAM crews claimed four Air Force Phantoms on 27 June, making it the campaign's single-costliest day since 10 May. Four airmen were killed and another four captured. In all, 21 U.S. aircraft were shot down that month, bringing Linebacker's total air losses to 41 by the end of June.[35]

Nevertheless, U.S. pilots were averaging 300 sorties per day by early July. And the effort was doing real damage. Vital infrastructure, industrial facilities, military installations, air defense assets, materiel caches, and more had been hit hard in every corner of the country. The campaign had been especially destructive to the country's transportation and logistics networks. Following two months of Linebacker, there were not more than 50 miles of uncut railroad track in the country, and some 400 bridges had been knocked out. This forced the North Vietnamese into the labor- intensive and time-consuming process of moving men, materiel, and supplies by truck and other means, often requiring lengthy detours around bombed-out bridge crossings and wrecked motorways. North Vietnam had some 20,000 trucks at its disposal and had shown remarkable resilience to U.S. interdiction efforts in the past. But several related factors further disrupted its transport and supply network. While concerted interdiction of truck traffic above the 20th parallel was deemed too risky for too little reward in the face of concentrated air defenses, mission planners instead opted for attacks on vehicle storage areas, repair facilities, and supply depots. There were no such restrictions in the country's southern panhandle, however, and fighter-bombers and B-52s attacked transport, supply, and communication lines with enthusiasm.[36]

But the resumed bombing of North Vietnam's heartland

allowed Hanoi to resurrect one of its more controversial claims from earlier in the war: that the U.S. was intentionally targeting the country's elaborate dike system to inflict flood, famine, and death. Hanoi's allegations focused on the Red River Delta, a flat, low-lying plain of 6,000 square miles encompassing Hanoi and Haiphong. Home to 15 million North Vietnamese in 1972, the Delta was situated at the confluence of several rivers and tributaries and was the country's primary rice-producing region. Historically subject to flooding, especially during the heavy rains of July and August, the region was protected by a complex water-control system of dams, locks, and some 2,500 miles of primary, secondary, and tertiary earthen dikes, some as wide as 200 feet at the flood line. But the system was far from perfect. In July 1971, heavier-than-expected rains had caused several dike breaches, resulting in the region's worst flooding in decades. Some 1.1 million acres were inundated, wiping out the area's rice crop and forcing the country to import food from the Soviet Union and China. The immense water pressure had significantly eroded the dikes, as well. Historically, government authorities would have directed ample resources to maintain and repair the system. But Hanoi had shifted the bulk of its labor force to the war effort, so sufficient repairs had not been made. In a rare government admission, the state-run newspaper *Hanoi Moi* reported in early summer 1972 that repairs had not met "technical requirements." In the same article, the mayor of Hanoi admitted that "dikes damaged by torrential rains in 1971 were never repaired."[37]

During Rolling Thunder, the Johnson administration had declared the dikes strictly off limits. But that did not stop communist leadership from waging a concerted misinformation campaign on foreign press and visitors. According to the party line, any bomb damage to its flood-control network was the result of deliberate U.S. attack. Meanwhile, Hanoi took advantage of Johnson's pledge by intentionally placing AAA guns, SAM sites, POL storage facilities, and other legitimate military targets directly on or adjacent to the dikes. The National Command Authorities faced a dilemma: let the targets stand or risk collateral damage to the dikes and the resultant outcry. The air-defense emplacements were particularly vexing. Allowing the weapons to go unmolested not only increased danger to U.S. pilots but decreased bombing accuracy, as well. Ironically, the latter often led to higher numbers of incidental civilian casualties. Nevertheless, Johnson and the NCA usually denied requests to hit targets on or near to the dikes.[38]

As Operation Linebacker heated up, so too did Hanoi's renewed propaganda campaign. The VWP politburo had identified what it believed to be "contradictions" between Nixon and the American public, and between the Republican and Democratic parties. By leveraging those tensions, the politburo hoped pressure from the antiwar movement and Nixon's Democratic opponents would damage his credibility at home and negotiating position in Paris.[39] By early May, Hanoi radio broadcasts were already accusing the U.S. of deliberately striking the "dike system in Namha Province" southeast of the capital. Dozens of similar accusations followed.[40] By late June, Xuan Thuy, one of North Vietnam's chief negotiators in Paris, had taken the charge to the world stage. He alleged the U.S. was deliberately bombing the retention dikes and "purposefully creating disaster for millions of people during the coming flood season." The allegations found a receptive audience. The accusation was picked up by *Agence France-Presse* Hanoi correspondent Jean Thoraval, who claimed on 11 July that he had personally witnessed U.S. planes targeting the dikes. Sweden's ambassador in Hanoi, Jean-Christophe Oberg, added his voice, alleging to have seen bomb-damaged dikes and that the attacks appeared "methodic." Several prominent world figures joined the growing chorus, including U.N. Secretary General Kurt Waldheim, who said he had received reports through "private unofficial channels" that the U.S. had been targeting the dikes. Waldheim declared that he was "deeply concerned" and called for an end to "this kind of bombing." Another was Eugene Carson Blake, general secretary of the World Council of Churches, who in a statement urged Nixon to stop bombing the dikes. And then there was actress and antiwar activist Jane Fonda.[41]

Then 34, Fonda had been an outspoken critic of the war for years. In March 1971, she toured with the FTA ("Fuck the Army") Show, an anti-military variety act that performed near U.S. bases at home and abroad. Intended as an explicit rebuke to actor Bob Hope's long-running, patriotically themed USO tour, the show featured prominent actors, comedians, musicians, and activists performing anti-war skits and musical numbers. The productions were filmed and released as the feature-length movie *F.T.A.* in late 1971.[42] The next summer, Fonda visited North Vietnam at the invitation of the North Vietnamese Committee of Solidarity with the American People. Clad in the white tunic and black pajama pants common among the peasantry, the actress arrived in early July, greeting her hosts from their "revolutionary comrades in America." Over the next two weeks, Fonda's North Vietnamese handlers led her to various sites purportedly damaged in U.S. airstrikes, irrigation dikes prominent among them. While in the country, Fonda voluntarily broadcast 10 live and taped propaganda statements aimed at American pilots and personnel. Among her many accusations, Fonda charged the U.S. government with lying and said it was intentionally targeting civilians. She urged pilots to "carefully consider what you are doing" before carrying out their orders. Seven American POWs from the Hanoi Hilton were eventually taken to a NVA film studio on the city's outskirts. There, as cameras rolled and guards stood watchful, they were questioned by Fonda during what was billed as a "press conference." One prisoner reportedly had to be tortured to force compliance.[43]

Fonda later claimed that the men told her they were being well treated, had not been tortured, and that, "without exception, they expressed shame at what they had done." Most infamously, the actress would at some point be filmed "manning" one of the city's antiaircraft guns while laughing with North Vietnamese soldiers and onlookers. Upon her return to the United States, Fonda presented 20 minutes of film footage at a New York press conference that purportedly showed bomb damage to various dikes. "I believe in my heart, profoundly, that the dikes are being bombed on purpose," Fonda said. For all this and more, the actress would become derisively known among many Vietnam veterans as "Hanoi Jane."[44]

The Nixon administration vehemently denied the accusations. "If it were the policy of the U.S. to bomb the dikes," said Nixon during a late-July press conference, "we would take them out, the significant part, in a week. We don't do so…because we are trying to avoid civilian casualties, not cause them."[45] Nixon may have been overstating the ability to destroy the dikes even if he had wanted to. Secret studies by the Air Force and CIA during Rolling Thunder had concluded that, because of its complexity and redundancy, the water control system would have been very difficult to significantly damage using airpower alone. Moreover, any nominal military benefits gained would be far outweighed by the worldwide outrage that would surely follow.[46]

Nevertheless, administration officials conceded it was possible some dikes had sustained "incidental" damage because of their proximity to legitimate military targets. Unlike Rolling Thunder, air commanders during Linebacker were authorized to hit targets near the dikes. But to minimize accidental structural damage pilots were restricted to "less destructive" anti- personnel weaponry such as guns, napalm, and cluster munitions. Beyond merely relying on official denials, however, the administration ordered the State Department, the

CIA, and the National Security Council (NSC) to examine the claims. The House Armed Services Committee (HASC) convened a formal inquiry, as well. Photographic evidence would be a crucial to the investigations.[47]

One of the men tasked with gathering such evidence was Lt. Craig Honour, an F-4J driver aboard the USS *Saratoga.* On the evening of 9 July, the VF-31 pilot and his shipmates were eagerly awaiting the next day's badly needed "standdown," a rare circumstance when all flight operations ceased for 24 hours. Crewmembers could instead enjoy a summer barbecue on the flight deck, see a movie, catch up on sleep, and so on. While liquor was officially off limits at sea, Honour had stashed a few cold beers for the occasion. But before he could crack his first clandestine brew, the squadron duty officer called down. No standdown for you, he was told. There was a special reconnaissance mission the next day, and Honour would fly escort. "Command authority needed pictures proving we were not bombing the dikes," he recalls ruefully. "So me and a poor photo guy had to risk our lives to go get them."[48]

The news went from bad to worse during the mission briefing. Not only would the pilots miss out on the ship's coveted standdown, but weather reports forecast the cloud cover at no more than 3,000 feet. Safety protocols prohibited flying under such conditions because it forced pilots to fly low—and straight into the sights of North Vietnamese antiaircraft gunners. "We had to fly below the clouds, and they knew that," says Honour. "Once they figured out our altitude, all they had to do was set their proximity fuses. We were sitting ducks." But the mission was "urgent," so safety standards were waived. Further, the *Saratoga* was the only flattop in the vicinity to carry the RA-5C Vigilante, the Navy's premier reconnaissance plane. Original designed as a nuclear bomber, the two-seater had been reequipped with sophisticated surveillance sensors and an advanced camera array that could gather high-resolution photos over a vast area. The RA-5 was also the "Ferrari" of long-range reconnaissance aircraft. Its sleek, clean airframe and powerful twin J79-10 engines made it fast and agile even at low altitudes. With no armaments to weigh it down, the Vigilante would be very difficult to keep up with.[49]

The next morning, a thick blanket of clouds hung at just 2,500 feet. "Great for taking pictures," says Honour, "horrible for survival." The recon area was in the heavily defended Red River Delta south of Haiphong, so the flight would have to skim the deck through a maelstrom of antiaircraft fire. To cut weight, Honour had ordered his Phantom stripped to the bone the night before, retaining only a centerline fuel tank and the air-to-air missiles he would need to ward off any lurking MiGs. Even so, he could make only about 850 miles per hour at that low altitude, still nearly a hundred slower than the RA-5. But a high-speed burn would be their only chance. "He told me, 'You've got to try to keep up with me,'" Honour recalls. And with that, the Vigilante pilot hit full afterburner and started his "racetrack" pattern over the recon area, cameras whirring. The circuit called for a series of highspeed straightaways and wide, left-arcing turns. "I'm on full afterburner, but I still can't keep up," says Honour. But the men had a plan. A few seconds before each left turn, Honour would cut hard inside the Vigilante, shortening his arc to momentarily close the distance. Meanwhile, glowing waves of creamy orange tracer fire snaked skyward, fiery shell blasts all around. Shrieking cockpit warnings alerted the men they were being swept by radar, but no SAMs came up to meet them. Honour, "sweating balls" because the Phantom's air conditioner did not work well at low altitudes, called a running commentary for the RA-5 pilot. "I'm telling him, 'We've got fire coming up, shells here, shells there.' It was insane. But we were going so fast they never could get a lock on us."[50]

By the time the recon flight went feet wet over the Tonkin Gulf, both pilots were bone dry on fuel. They had run full afterburner the whole way. Honour says that it was only later that he began to fully grasp why he had been put in harm's way. While the North's propaganda effort had begun well before her July visit, the pilot still places the blame squarely on Fonda and her efforts to add fuel to the fire. "I realized what a pawn I'd been," he says. "I had to risk my life because that idiot decided to sit on a dike and say we were deliberately bombing them. I don't have a good feeling about her, to say the least. May she rot."[51]

Throughout the summer, other reconnaissance aircrews would be ordered to lay their lives on the line to refute Hanoi's claims. Thanks to their efforts, the CIA, State Department, NSC, and HASC investigations had by the end of August produced copious evidence showing that the dikes were not being intentionally targeted. Any damage done, according to the reports, had been incidental and likely resulted from proximity to military targets— including those intentionally positioned by the North Vietnamese. Even prominent Vietnam War opponents, including past and current government officials with close knowledge of U.S. policy, doubted the veracity of Hanoi's claims. "I know for a fact that the target lists do not include dikes," one well-informed former official said. "If we wanted to bomb them, we would." There was, according to the official, "too little to gain from such a risk."[52] Joseph Kraft, a syndicated columnist and outspoken critic of American involvement in Vietnam, returned from his own fact-finding tour of North Vietnam in August. The newsman observed that any bomb damage to the dikes seemed "harum-scarum" and not "methodical," and that there appeared to be "no deliberate American drive to bomb the dikes." Kraft did conclude, however, that since so many legitimate targets like roads, railroad tracks, and bridges ran parallel to the flood-control system, some accidental damage was inevitable…as were Hanoi's attempts to exploit it.[53] Still, as warfare legal scholar W. Hays Parks wrote, "the United States was able to rebut these charges only through substantial expenditure of valuable national assets and risk of life by U.S. military personnel. Thousands of man-hours were diverted from the war effort to respond to the allegations, raising a question for future military operations as to the degree to which the military must plan to respond to spurious charges."[54]

§

Back on the coast, the U.S. import blockade continued to pay dividends. The closure of the North's ports and the resulting reliance on overland transport from China had drastically slashed imports from their pre-May levels of about 150,000 tons to just 30,000 tons monthly. Among the critical supplies hardest hit were desperately needed replacement SA-2 missiles, hundreds of which had been barrage fired in the early days of Linebacker. Despite the increasing number and ferocity of Linebacker strikes, SAM launches had fallen to about 10 per day by early June, indicating stocks were running low. Also in urgent need of replenishment were the spare parts and other equipment needed to repair or replace damaged and destroyed SA-2 launchers and radar systems.[55]

Both services continued to place a high priority on degrading SAM capabilities, with each strike mission accompanied by either Navy Iron Hands or Air Force Wild Weasels. Especially effective had been the Air Force's "hunter-killer" (H-K) flights. The tactic called for two F-4Es equipped with cluster munitions to work in tandem with a pair of F-105G Wild Weasels armed with radar- homing missiles. This usually meant either the anti-radiation AGM- 78 Standard ARM or the AGM-45 Shrike. Aircrews overwhelmingly preferred the Standard, which was faster, had a larger warhead, and boasted a range about four times that of the Shrike. But chronic Air Forces shortages—not least because of the Standard's much larger size and heftier price tag—usually meant Wild Weasel crews were stuck with the AGM-45. Regardless, the aim of hunter-killer teams was to destroy not just the SAM's guidance radar but to wipe out the rest of the site with MK-20 cluster bombs. MiG pilots usually looked to target the more vulnerable F-105s, so the Phantoms also acted as escorts for the Thuds. Anti-air defense suppression was dangerous work. The Weasels were usually first in and last out, clearing a path for the strike element by flying straight into the maw of enemy antiaircraft fire. There they fought a deadly duel with SA-2 crews and AAA gunners while lurking enemy fighters waited to pounce. Their motto, YGBSM ("You Gotta Be Shitting Me!"), told the tale. Some of the war's most storied aviators hailed from Weasel ranks. Capt. Merlyn Dethlefsen and Maj. Leo Thorsness would go on to earn the Medal of Honor, while 15 others would be awarded the Air Force Cross. And with good reason. Life as a Weasel meant a daily dance with death. By war's end, 48 Wild Weasels would be shot down, with 35 air crewmembers killed or missing, and another 17 taken prisoner.[56]

Life could be just as dangerous for Weasel groundcrews. On 17 May 1972, Sgt. Edward "Eddie" Eiler was on the job at Korat Royal Thai Air Force Base when he heard something that did not sound right. At about 0700, the Air Force jet mechanic looked up from behind the concrete revetment where he was working and spied a 561st TFS Wild Weasel skidding wildly down the runway on its belly. The F-105's left main landing gear had snapped upon touchdown, and the Thud was throwing a cascade of yellow-orange sparks. One of its 450-gallon fuel tanks was already ablaze. As the stricken craft slid to a halt, fire-rescuers Staff Sgt. Raymond M. Daubendiek, 32 and Sgt. James Lathon, 23, along with others from the 388th Combat Engineer

Squadron, pushed past the growing wall of flames to reach the crew before the tank exploded. Frantic seconds ticked by as the rescuers struggled to free pilot Maj. Donald Kilgus and WSO Maj. James Dozier, the intense heat now nearly unbearable. Daubendiek and Lathon at last pulled the aviators free, helping them a safe distance from the plane. Then they went back to help fight the fire. More minutes passed as fire-rescue crews seemed to be getting a handle on the inferno. "Then I saw a little black trail of smoke coming up from the aircraft," Eiler says. "All of a sudden, there was this tremendous explosion and bodies were flying through the air. It was just awful."[57]

It had been a Shrike, not fuel tank, that blew. Kilgus and Dozier's early-morning mission had seen no action, so the aircrew had returned still armed with missiles. The intense heat had detonated one of the 150-pound warheads, spraying flaming shrapnel in all directions. Daubendiek and Lathon, along with four Thai ground crewmen, were killed instantly. A dozen others were wounded, some losing limbs or otherwise horribly maimed. The fire still raging, Eiler watched in awe as an APC careered out of nowhere toward the inferno, a rocket lying on the grass near the blaze. If it exploded, thought Eiler, more lives would surely be lost. And then yet another act of courage that day. "The guy backed up to this thing…put the door down, and this brave person got out, tied a cable to the rocket, and towed it away. I'll never forget this."[58]

§

The blockade was also staunching the flow of POL products from the Soviets and other Eastern Bloc nations. POL was the lifeblood upon which the North's mechanized invasion force depended. U.S. strikes on rail lines from China had further cut POL imports to a trickle, and air attacks on the North's storage facilities had shrunk available stores to dangerously low levels. Likewise, anti-aircraft and other ammunition stocks were falling to critical levels. Contemporary U.S. estimates calculated that Linebacker and Pocket Money had already cut resources to NVA divisions in the South by 20 percent. By July, the invasion force was already experiencing marked shortages in food, fuel, and ammunition.[59]

But the relative success of Linebacker had been costly. Over the first two months of the campaign, 41 U.S. aircraft had been lost to ground fire and air combat.[60] Air-to-air engagements had taken an especially heavy toll on the Air Force. The service had managed just a 1:1 kill ratio during Linebacker's early months. This would worsen toward the end of June and into early July. Over a 12-day period, the Air Force lost seven straight fighters without claiming a single MiG in return. By the end of the month, the service had lost 13 F-4s to air combat while downing just seven enemy fighters.[61]

Several factors contributed to this alarming trend. First, the performance record of both the AIM-7 Sparrow and the AIM-9E variant of the Sidewinder had been atrocious. The Sparrow alone hit just 20 percent of its targets. Aside from various design and mechanical flaws, both missiles had been conceived as beyond visual range (BVR) weapons for a superpower clash with the Soviet Union. While that conflict never materialized, the missiles did prove woefully inadequate to the close-in dogfighting of the Vietnam War. As the summer wore on, wider distribution of the AIM-7E-2 variant, the so-called "dogfight Sparrow," helped somewhat since it was more effective at shorter ranges. But it was hardly a game-changer.[62]

The situation improved a bit more as greater numbers of Air Force F-4Es arrived in theater. The variant offered several advantages over the Phantom D, including a more powerful J79-GE- 17 twin-turbojet engine and leading-edge wing slats that improved low-speed maneuverability. The aircraft was also equipped with a 20 mm M61A1 Vulcan rotary cannon. Housed internally and mounted in the nose, the six-barreled Vulcan could spit 6,000 rounds a minute, providing air crews with a lethal dogfighting weapon. By the end of Linebacker, Air Force F-4E gun kills would account for five MiG shootdowns.[63]

But perhaps no change better improved the service's fortunes than ditching its outdated "Fluid Four" formation in favor of the Navy's more flexible "Loose Deuce." Indeed, the Navy had achieved a 6:1 kill ratio against enemy fighters during May and June, a trend that would continue throughout Linebacker.[64] The Loose Deuce called for a flight element of two Phantoms to fly parallel at a distance of about one mile. This enabled crews to mutually support one another should an adversary gain a rear-quarter firing position. When on the attack, either fighter could assume the role of tactical lead, while the other covered his partner's six o'clock. Depending on how the fight evolved, the roles might change several times and usually depended on which pilot was in the best position to shoot.[65]

By contrast, the Air Force's Fluid Four was a holdover from the propeller-driven dogfights of World War II. The formation featured a four-ship flight divided into two equal elements of lead and wingman. Unlike the Loose Deuce, where either aircraft could shoot, Fluid Four protocols authorized only element leads to fire, theoretically cutting a flight's firepower in half. Instead, wingmen acted as little more than escorts and were obliged to fly a rigid "welded wing" pattern requiring them to trail element leads by no more than 2,000 feet. This constraint required intense concentration to avoid crashing into the leads, sometimes hindering aircrews from identifying threats as they emerged. Ultimately, the Loose Deuce would enable Navy and eventually Air Force aircrews to respond more quickly and effectively to air combat over North Vietnam. All of the above, including growing aircrew experience and expertise as the campaign wore on, helped the Air Force rebound to a 4:1 kill ratio by the end of the summer.[66]

As the Air Force and Navy worked to crush North Vietnam's warmaking capability and force its leaders back to the bargaining table, allied forces south of the 17th parallel struggled to halt Hanoi's Nguyen Hue Offensive. Some 200,000 North Vietnamese troops, backed by hundreds of tanks and heavy artillery pieces, continued to grind South Vietnam on three fronts. Nixon, meanwhile, determined to turn the war over to the South Vietnamese, pushed his Vietnamization policy ahead at full speed. The few American ground combat units still in-country, like the 196th Infantry Brigade at Da Nang Airbase, were there only to safeguard the remaining handful of U.S. military installations. By 11 August, 3rd Battalion, 21st Infantry and Battery G, 29th Field Artillery Regiment became the last U.S. combat units to leave Vietnam.[67] The president further announced that U.S. personnel in South Vietnam would again plummet to just 27,000 by 1 December, the lowest level since 1964.[68] For the first time in nearly a decade, South Vietnam's armed forces would be primarily responsible for defending their country. But they were not alone just yet. U.S. leaders were determined to put American airpower—including waves of B-52 heavy bombers—to devastating effect to roll back Hanoi's Nguyen Hue campaign and preserve their South Vietnamese ally.

6

ROLL BACK

May - October

By 9 May, NVA forces laying siege to the Binh Long provincial capital of An Loc in III Corps were ready to give it another go. Since the initial failure to capture the city in the third week of April, encircling communist forces had poured in three relatively fresh regiments from the 5th NVA Division to bolster the coming assault. Two additional regiments were pulled from the 7th Division, while what was left of the 9th Division, badly mauled during the first attempt to take the city, rounded out the ranks. All available armor and artillery were marshaled for the effort. In all, the North Vietnamese had approximately 20,000 troops for the assault. The man in charge of holding the city was 5th ARVN Division commander Brig. Gen. Le Van Hung, the same who had weathered the first attempt. Though the city had remained cut off and completely dependent upon U.S. aerial resupply since early April, Hung had—with help from U.S. 1st Cavalry Division airlift— managed to bring in a few reinforcements of his own. To the 4,000 or so survivors of the 5th Division and the remnants of the territorial militia, the 1st Cav's helicopters added elements of the 3rd Ranger Group and a battalion of airborne Rangers. Still, the general's force could not have numbered more than 7,000 troops, many hungry and wounded from more than a month of fighting. With fewer than a third of what opposed him, Hung ordered his weary men to dig trenchworks and construct bunkers all along his lines. He hoped it would be enough.[1]

As darkness settled on 10 May, the NVA let loose with a blistering artillery barrage, raining some 8,000 shells throughout the night and early morning. The 5th NVA Division then led the way, as infantry and armor pushed out from enclaves on the city's north end. South Vietnamese Airborne Rangers, backed once more by M72 light anti-tank weapons and a torrent of deadly accurate 20 mm rotary cannon fire from orbiting AC-130 Spectre gunships, inflicted heavy casualties. But only after a fusillade of close air support by U.S. and RVNAF fighter-bombers did the NVA assault finally grind to a stop. Once stationary and in the open, the communists were even more ripe for tactical air strikes. Allied airpower, under the skilled direction of orbiting American and South Vietnamese Forward Air Controllers (FACs), drowned the enemy spearhead in a deluge of some 300 strike sorties on the first day alone.[2]

Meanwhile, troops of the 9th NVA Division drove in from their footholds on the west end, part of a plan to split the city's defenders and chop them up piecemeal. Manning the western ramparts was the 7th ARVN Regiment, whose infantrymen put up enough of a fight to slow the onrush, allowing allied airpower to once more take its deadly toll. Next came a continuous train of B-52 Arc Light strikes, three-plane cells saturating enemy positions as close as 600 yards west of the city. The strikes rumbled every hour on the hour for a full day, crushing the NVA assault under thousands of tons of high explosives. Again, as had happened since the earliest days of Nguyen Hue, North Vietnamese planners had not accounted for what allied air strikes could do to their massed formations. NVA foot soldiers were now paying the terrible price of that miscalculation. As dawn broke on 12 May, daylight told the tale. Hundreds of NVA corpses littered the ground in every direction, while scores of North Vietnamese tanks and other vehicles lay smoldering and wrecked.[3]

Marine Corps A-4 Skyhawk pilots had done a lot of the damage. On 17 May, 32 of the venerable attack bombers of Marine Aircraft Group 12 (MAG-12) under Col. Dean C. Macho had relocated from Iwakuni, Japan to Bien Hoa Airbase just north of Saigon. Two days later, Marines with the VMA-211 Avengers and VMA-311 Tomcats were already flying close strike missions, most of them in support of South Vietnamese forces defending An Loc.[4]

As the fight raged below, two of those pilots, VMA-311's Lt. Sebastian "Vince" Massimini and wingman Bill Peters, were orbiting nearby when the excited voice of an Air Force FAC crackled over their headsets. 'Tanks!' came the call. The Tomcats' A-4s were armed with eight unguided Zuni rockets and six 750-pound CBU cluster bombs. "Perfect if you found a lot of guys in the open…[but] cluster munitions are not what you want for tanks," says Massimini. Still, a few well-placed Zunis might do the trick. The men started their "racetrack" pattern, each circling the target in opposite directions. To put ordnance on the mark, the A-4's manual gun sight required a steep dive angle and intense focus—without concern for enemy fire. "You can't concentrate on antiaircraft and also hit the target," Massimini says. "That's just life on an attack airplane." So, while one pilot dove, the other circled above to call out SAM launches or AAA that might be getting too close for comfort. And there was a lot to watch for. The ZSU-23 was the weapon pilots feared most. The twin-barreled 23 mm always seemed in ample supply, and today was no exception, as glowing ribbons of green tracer fire snaked skyward. [5]

Massimini dipped his nose and dove, fighting gravity—and fear—to keep the Skyhawk's gunsight on target. He let loose with a flurry of rockets, churning the earth all around the tanks but scoring no hits. Peters tried his luck, but he too hit nothing but dirt. As the tanks lurched toward the life-saving cover of the tree line, it was now or never. All that was left were the cluster munitions. Worth a try. So the pilots peppered the rumbling tanks with thousands of bomblets from the CBUs. The T-55's armor is thick just about everywhere except for a small engine exhaust port behind the turret. "And don't you know those bomblets hit the backs of those tanks a caught both on fire!" We're out there with the wrong armaments…but you know, it worked!" Massimini says, laughing.[6]

MAG-12's operations would eventually expand to include attack missions on communist staging areas in Cambodia, tactical strikes in IV Corps, and close air support for Kontum's defense. By virtue of aerial refueling, group A-4s would even assist MAG-15 in defending Hue in northern I Corps. By year's end, MAG-12 would fly 12,574 combat sorties and drop some 18,903 tons of ordnance. On 26 January, the group became the last fixed-wing American aviation unit to leave Vietnam. But the effort had not been without cost. MAG-12 would lose three Skyhawks during its eight-month combat tour, with three Marines killed and another 11 wounded. Another A-4 pilot, Capt. James P. "Waldo" Walsh of VMA-211, would be shot down and listed as missing in action on 26 September. Well-liked by all, Walsh's loss cast a pall over MAG-12.[7]

Only after the war did his groupmates learn of their friend's fate. On 29 January, more than a dozen Avengers and Tomcats had gathered in the Cubi Point officer's club in the Philippines to celebrate the end of the war. "There were a lot of drinks flowing, and let's just say we were feeling very good," Massimini says with a chuckle. Suddenly, another pilot burst in waving the latest issue of Stars and Stripes. 'Waldo's on the list!' he screamed. The men, many of whom had ordered POW bracelets in Walsh's honor, excitedly scanned the page and found their friend's name among the prisoners soon to be released by the Viet Cong. Walsh held the dubious distinction of being the last Marine captured in the Vietnam War. "It had been eight months and that was the first time we knew what had happened to him," Massimini says, his voice breaking. "He was in a bamboo cage that whole time in South Vietnam. There were literally tears in everybody's eyes."[8]

Over the next few weeks, ARVN defenders weathered several more truncated NVA assaults on An Loc but turned back each foray. The near-continuous airstrikes on NVA formations had also destroyed many enemy anti-aircraft emplacements, finally allowing an unmolested flow of aerial resupply. Over several days, some 1,000 seriously wounded were evacuated, while nearly 1,700 fresh troops were choppered in to replace them. By the end of the month, South Vietnamese forces had widened their defensive perimeter and cleared the remaining pockets of NVA resistance from the center's city. By 12 June, the final enemy enclaves in An Loc's northern and western outskirts had been either destroyed or driven out. By 18 June, III Corps commander Gen. Lt. Gen. Nguyen Van Minh declared the siege officially lifted, though sporadic fighting and shelling continued for weeks.[9]

Hung's 5th ARVN Division, which had shouldered much of the burden, had lost nearly a third of its men killed and wounded. Minh ordered the 18th Division to replace it. After a massive helicopter lift, the division assumed responsibility for An Loc's defense on 13 June. Finally, Gen. Minh pulled the exhausted and depleted 21st ARVN Division, which had suffered some 600 killed and more than 3,000 wounded during the weekslong fight to clear the NVA blocking positions south on Highway 13. The newly arrived 25th Division immediately set about encircling and destroying this last vestige of enemy resistance, finally clearing it on 20 July. With the road to An

Loc now open, the 66-day siege was truly over.[10]

Since the attack had begun in mid-April, the fight to hold An Loc had cost the South Vietnamese 5,400 casualties, at least 2,300 of whom were killed or missing in action. Three months of near-continuous shelling—an estimated 78,000 rounds—had reduced the city to a blasted hellscape, crumbled and smoldering rubble in every direction. As is so often the case in war, civilians bore a heavy cost. Thousands had been killed or were missing, and thousands more made refugees.[11]

The tally was far worse on the other side. Since their opening moves on 7 April, three of North Vietnam's best divisions had lost a combined 10,000 dead and 15,000 wounded. Nearly 100 T-54/55s and other mechanized assets lay wrecked and smoldering on the battlefield. Under the command of Brig. Gen. Le Van Hung, the city's tenacious defenders, backed by prodigious American and South Vietnamese airpower, had cut through the enemy ranks like a murderous scythe. The U.S. and RVNAF had flown some 9,200 close air support sorties and another 262 B-52 Arc Light missions against NVA positions at An Loc and along Highway 13. Such was the carnage, the highway became known as the "Road of Death" among communist soldiers. The bid to conquer An Loc and establish a "liberation" PRG authority just 50 miles northwest of Saigon had come to ruin—along with Le Duan's hopes of a civilian uprising. President Nguyen Van Thieu, who flew in on 7 July to personally congratulate the defenders, dubbed Hung the "hero of An Loc." Meanwhile, the Nixon administration touted the victory as proof that its Vietnamization policy was working. Of course, the success at An Loc would not have been possible without the herculean efforts of U.S. airlift crews, ferocious B-52 Arc Light missions, and an unrelenting torrent of American-directed close airstrikes. All of which reinforced the ongoing concern with

Vietnamization: South Vietnam's continuing dependence on U.S. military know-how and airpower to stave off defeat. Still, the victory at An Loc helped break the back of the Nguyen Hue Offensive in III Corps, and for the moment, that was enough.[12]

Despite sporadic jabs and thrusts by both sides over the summer, nothing much changed in Binh Long Province. Saigon, dealing with crises in I Corps and in the Central Highlands, was content to solidify control over An Loc, the areas immediately north and east of the city, and Highway 13 south toward Saigon. A few limited attempts to push north up the highway toward the captured district capital of Loc Ninh, which had fallen to the NVA in early April, produced very little. Government forces in III Corps simply lacked the men and materiel after months of hard fighting. By September, the lines had mostly stabilized, with the NVA now controlling a chunk of northern Tay Ninh and Binh Long provinces, while Saigon retained dominion over lands to the south. This standoff would maintain for the rest of the year.[13]

§

About 300 miles to the northeast, another vital South Vietnamese provincial capital, this time in the West-Central Highlands of II Corps, stood in danger of being overrun. Kontum City represented the seat of government power in the region, and its capture would not only ensure the loss of the rest of the province but threaten equally important Pleiku City just 30 miles south. Since late April, some 18,000 troops of the 320th NVA Division, the 2nd Division, armor from the 203rd Tank Regiment, and assorted sapper units had seized control of northwestern Kontum Province before pushing south to gradually coil around the city. The encirclement was complete in early May when the 95B NVA Regiment established a strong blocking position in the

Chu Pao Pass on Highway 14 to the south, cutting the city off from its only source of reinforcement and resupply at Pleiku. Not that there was much to offer. As Saigon desperately sought to counter NVA gains throughout the country, its general force reserves were stretched to the breaking point. Kontum was on its own.[14]

Opposing the NVA was Col. Ly Tong Ba, 41, and the three regiments of his 23rd ARVN Division, a few squadrons of M-41 Walker Bulldog light tanks from the 8th Cavalry Regiment, along with a few hundred Regional and Popular Forces territorial militiamen, colloquially known among American 2nd Regional Assistance Command (RAC) advisers as "Ruff-Puffs." Ba, who had commanded the 7th ARVN Division during 1966's Operation Lam Son II, intuited that the main NVA attack would roll south down Highway 14 and hit the city from the north. The colonel established a defensive perimeter of trenchworks just north and northwest of the city, anchoring the line with three regimental strongpoints in the shape of an "L." The 44th Regiment sat astride Route 41 about two and half miles northwest of the city, with the 45th regiment to its rear protecting the northwestern edges of the city. Meanwhile, the 53rd regiment screened the city's northeastern approaches, including Kontum airfield. Ba placed the RF-PF militia behind a parapet system to guard the city's southern approaches. Over the preceding weeks, the colonel had personally overseen the construction of defensive positions, particularly ensuring that they had ample overhead cover. Ba also soon became known for moving about the trench lines to offer guidance and encouragement to his junior officers and infantrymen. Though untested, the men of the 23rd responded to their commander's attentiveness, bolstering morale for the fight ahead. They would soon need it.[15]

The NVA struck at dawn on 14 May. As expected, the main attack rolled south along Highway 14, as the 48th and 64th regiments of the 320th Division, spearheaded by dozens of T-54/55 tanks, drove straight for the city. Meanwhile, the 28th NVA regiment hit the 44th ARVN to the northwest, while the 1st Regiment, 2nd NVA Division struck the 53rd ARVN Regiment's lines near the airfield. Within an hour, all three of Ba's regiments were fixed and heavily engaged. Meanwhile, the 141st Regiment, 2nd NVA Division, along with sapper units, had looped around to press the territorial militia's defenses to the south. But the NVA's route of advance had left it exposed. At Ba's direction, South Vietnamese gunners had pre- sighted coordinates all along the highway, enabling them to respond with a ferocious—and highly accurate—artillery barrage, saturating the onrushing NVA assault with hundreds of 105 mm and 155 mm high explosive rounds. The NVA hit back with their own heavy guns and rocket attacks, but the ARVN positions had been well constructed. Meanwhile, South Vietnamese infantry armed with M72 light anti-tank weapons began to take a deadly toll on NVA armor. The communist attack soon faltered under the combined weight, trapping the men in the open. On cue, U.S. Cobra gunships and RVNAF A-37 Dragonfly tactical bombers roared in to provide close air support. Coordinated by American advisers, CAS pummeled the enemy formations bunching up to the front of ARVN lines. A few UH-1B helicopter gunships were on hand, as well. Outfitted with the new XM26 TOW (tracked, optically wire-guided) anti-tank missile, the Hueys soon knocked out several T-55s as they desperately sought cover. The TOW would see extensive use during the siege, repeatedly proving its worth as a formidable anti-tank weapon. The NVA assault soon petered out, though sporadic probing attacks continued throughout the day.[16]

Under cover of darkness, however, the communists launched a second push to crack ARVN lines. And it looked at first like they might succeed. A strong attack east of Highway 14 drove a wedge between the 44th and 53rd regiments. Waves of NVA infantry poured in to widen the breach and threaten the ARVN flanks. If they succeeded, Ba's isolated regiments would be chopped up and destroyed in detail. The situation desperate, Ba had one last ace up his sleeve. He first pulled his men back to form a shorter perimeter. Then he ordered the two B-52 Arc Light strikes planned for that night to shift to the NVA breach. The darkness soon exploded in a terrifying cacophony of blinding orange fireballs and thunderous explosions, most just a few hundred yards from friendly lines. The earth shook under thousands of tons of high explosives as the NVA salient transformed into an abattoir. Hundreds of enemy infantrymen were simply wiped out.[17]

The next day, II Corps commander Gen. Nguyen Van Toan risked a flight into the beleaguered city to offer his personal congratulations. The general was accompanied by John Paul Vann, a near legend among old Vietnam hands. As an Army lieutenant colonel in early 1962, Vann had volunteered to serve as senior adviser to IV Corps, a region encompassing the Mekong Delta and one that had experienced a virulent communist insurgency during the war's early days. The Korean War veteran and former Army Ranger gradually became a sharp critic of both the war's conduct and of then-President Ngo Dinh Diem, whom the American viewed as hopelessly corrupt. Vann became convinced that only Diem's ouster, followed by dramatic political reforms in Saigon, could prevent a communist takeover of South Vietnam. When Washington continued to support Diem into 1963, Vann resigned in protest and signed on with the civilian U.S. Agency for International Development, a governmental organization engaged in rural pacification. There, Vann worked to advance the cause of rural reform in an effort to win the "hearts and minds" of the population. His tenacity and personal bravery soon earned him the admiration of Vietnamese and Americans alike. In May 1971, Vann was named senior adviser for II Corps, a post usually reserved for a two-star general. Since then, he had helped guide the region through a remarkable period which saw communist influence in the backcountry dramatically wane while Saigon's authority continued to spread. Vann had also used his considerable experience and force of will to help II Corps survive the NVA's recent invasion of the Central Highlands front. Vann, along with a team of American 2nd RAC advisers, would remain in Kontum throughout the siege.[18]

And the South Vietnamese colonel would need all the help he could get. The NVA were still invested in strength all about the city and would soon come again. Likewise, the previous day's battle had taken its own deadly toll on Ba's three regiments. He simply did not enough men to hold the original perimeter. So the division commander ordered his troops back once more, closer to the city's outskirts and into a more-defensible line. He then pulled the 44th Regiment off the line to serve as a reserve element. The ARVN defenders did not have long to wait. Over the next several days, the NVA doused the city in near-continuous artillery and rocket fire, pummeling South Vietnamese positions and forcing the closure of the airstrip. Meanwhile, the 2nd NVA Division launched repeated assaults on the center of the ARVN line. By 19 May, the enemy had again succeeded in driving a perilous wedge between the 45th and 53rd regiments. Ba ordered in his reserve, backed by a squadron of Walker Bulldogs, to plug the hole. They barely succeeded. Vann coordinated a flurry of CAS strikes to once more save the day, as Cobra attack choppers and South Vietnamese AC-47 Spooky gunships drowned the communist assault in a deluge of fiery ordnance. Meanwhile, thunderous B-52 strikes against staging and logistics areas in the enemy's rear stymied efforts to resupply forward troops or bring up reinforcements to exploit the gap. By the night of 21 May, ARVN's lines had again held, and another 1,000 NVA soldiers lay dead and dying in the killing fields before Kontum.[19]

On the same day, Gen. Toan launched an ill-fated attempt to

dislodge the NVA blocking force at the Chu Pao Pass in hopes of reopening Highway 14 to the south. Despite heavy bombardment by tactical strike aircraft, the enemy position was too well entrenched. Both South Vietnamese Ranger groups charged with the mission were thrown back with heavy losses. Kontum would remain cut off. After a three-day lull in which both sides attempted to regroup, the NVA finally let loose with its most intense artillery bombardment on 24 May. Another regimental-sized assault backed by two sapper battalions finally succeeded in breaching the RF-PF defenses in the south, with communist troops gaining their first firm foothold in the city's outskirts. At the same time, a heavy infantry assault on the northeastern edge of the city, spearheaded by remaining NVA armor, pressed the beleaguered remnants of Ba's 53rd Regiment to their breaking point. Meanwhile, the enemy had edged his big guns— including captured American-supplied 155 mm howitzers—ever closer to the city. Soon, they were in range of South Vietnamese artillery, which NVA gunners proceeded to systematically destroy with murderously accurate counterbattery fire. By the next day, all ARVN artillery had been knocked out, leaving the city at the full mercy of North Vietnamese canon. Vann called upon Gen. Creighton Abrams, MACV commanding officer, to send every available aircraft to Kontum's defense. A swarm of U.S. attack fighters converged from all corners, devastating communist assault waves and turning back the attack in the northeastern sector. But the NVA remained firmly ensconced in Kontum's southern sector.[20]

Over the next two days, the communists threw waves of infantry and armor against the center of ARNV's defenses in the north, finally cutting off the 45th Regiment from the city on the left. Sensing opportunity, NVA commanders poured in reserves to widen the gap. Only a desperate, close-quarters stand by exhausted South Vietnamese defenders, backed by ferocious, danger-close tactical airstrikes, prevented ARVN's forward positions from being overrun. Throughout the night, NVA assault teams doggedly worked their way deeper into the northern outskirts. By the afternoon of 27 May, Ba had seen enough. His lines thin and faltering, the division commander reluctantly constricted his perimeter once more, this time into the city itself. NVA commanders were facing their own dire circumstances. They had already sustained horrendous casualties over a week of heavy fighting, while unrelenting B-52 strikes had devastated their supply trains and lines of communication. Further, the drenching monsoon season over the Highlands had begun, turning pathways to muck and making resupply and reinforcement even more difficult. If the battle of Kontum were to be won, it would have to be now or never.[21]

In the predawn hours of 28 May, the NVA unleashed their last all-out push. Heavy artillery blanketed the city, while commanders hurled their remaining infantry and armor into the heart of the South Vietnamese defense. Once again, it would not be enough. Desperate fighting by ARVN troops, themselves drained and low on ammunition, paired with round after round of vicious tactical airstrikes, finally broke the communist attack once and for all. Meanwhile, a savage house-to-house counterattack by RF-PF militia succeeded in pushing the NVA from the city's southern sector. By the afternoon of 30 May, South Vietnamese reconnaissance patrols reported the remnants of the NVA's assault force slinking away to the northwest. That afternoon, President Thieu choppered into Kontum to personally congratulate Ba and promote him to brigadier general. Mopping up operations would continue for another week until the siege of Kontum was declared officially over on 6 June.[22]

As for John Paul Vann, he would die in a helicopter crash on 9 June just days after helping clear the last pockets of enemy resistance near the city. He was 47. Vann was buried a week later at Arlington

National Cemetery. Among those in attendance were Army Chief of Staff and former MACV commander Gen. William Westmoreland and USAF Maj. Gen. Edward Lansdale (Ret.), who had played a pivotal role in helping shape early U.S. policy in Vietnam.[23]

The Central Highlands front of Hanoi's Nguyen Hue campaign had so far cost the North tens of thousands of dead and wounded in just under two months of fighting. At least 100 tanks and numerous artillery pieces, increasingly difficult to replace in the face of the continuing U.S. blockade, were also lost. About 4,000 alone had been killed in the failed bid to capture Kontum City. Shocked and exhausted by their manpower and materiel losses, the NVA pulled back to regroup and refit. Remnants of the 320th Division retreated to the relative safety of their recently conquered territory at Canh and Dak To II in northwestern Kontum Province. Meanwhile, what was left of the 2nd Division slipped across the border and into the sheltering jungles of neighboring Quang Ngai Province to the east.[24]

The South Vietnamese, themselves ravaged by months of hard bloody fighting, were in no hurry to pursue. Despite their defeat at Kontum, NVA troops still controlled provincial areas to the west. II Corps commander Gen. Toan wanted them pushed out. But Saigon, itself grappling with manpower and equipment shortages throughout the country, could spare little to properly reinforce and reequip Brig. Gen. Ba's 23rd Division for the effort. Other than a push up Highway 14 toward Vo Dinh, and a few airmobile operations into the Tan Canh area to keep the enemy off balance, ARVN activity in the province was decidedly muted. The effort in late June to clear the NVA's blocking position south of the city at Chu Poa Pass, however, was much more successful. After more hard fighting and casualties, South Vietnamese forces finally dislodged the communists on 26 June. Highway 14 between Kontum and Pleiku was officially reopened on 30 June, though commercial traffic continued to experience sporadic harassing attacks over coming months. As summer turned to fall, the new territorial lines had remained largely intact. North Vietnamese troops dominated the sparsely populated northwestern reaches of Kontum Province, including the all-important border regions with Laos and Cambodia. This ensured the unimpeded infiltration of men and materiel from the Ho Chi Minh Trail. For its part, South Vietnam maintained control over most of Highway 14, Kontum City, and the district capitals, along with the bulk of the civilian population.[25]

Meanwhile, the reconstituted 22nd ARVN Division, so devastated during the NVA's early April push into the Central Highlands, was now operating in Binh Dinh Province to the east. Under its new commander, Brig. Gen. Phan Dinh Niem, the division first turned back enemy efforts to seize the entire province in early June, which would have effectively split South Vietnam in two and cut off the five provinces of I Corps to the north. By mid-July, the division, along with the 6th Ranger Group, had gone over to the attack, recapturing several provincial towns and reopening Highway 1 north to the Quang Ngai border. By August, zones of control in Binh Dinh had reverted to their pre-invasion status quo. Government forces dominated the coastal population centers and Highway 1, while the NVA continued to occupy the mountainous and remote An Loa Valley to the west—the same region they had used to launch their attacks in April. As with Binh Long Province, the battlelines in Kontum and Binh Dinh would remain largely unchanged for the remainder of 1972.[26]

§

Even as the situation stabilized on the III Corps and Central Highlands fronts, months of hard fighting lay ahead in northern I

Corps. After the early May debacle that saw both the loss of Quang Tri Province and Hue City all-but encircled, President Nguyen Van Thieu recalled corps commander Lt. Gen. Hoang Xuan Lam and replaced him with Lt. Gen. Ngo Quang Truong, the aggressive and respected IV Corps commander. One of Truong's first official acts was to move his command headquarters into Hue where he could assume direct control over its defense. But the general had his work cut out. I Corps' northern command had been gutted after losing the entire 3rd ARVN division, hundreds of armored vehicles, and nearly all of the military region's heavy artillery during the blistering North Vietnamese attacks in April. Truong set about cobbling together a coherent and effective fighting force until Saigon could free up reinforcements. He unified under a single divisional command the remnants of three Marine brigades, badly mauled during the previous month. He tasked the new division with holding the line north of Hue, while the 1st ARVN Division, which had been forced to pull back in the face of a heavy push by NVA divisions out of the A Shau Valley, would defend to the west and south. A few Regional Force territorial militia units filled out Truong's rather meager order of battle. The situation improved somewhat when Saigon released the 2nd Airborne Brigade. With its arrival on 8 May, Truong's troop strength now sat at nearly 30,000 men. For now, it would have to be enough. Rather than remain in static defensive positions, however, the general looked to push the enemy back.[27]

Over the next several weeks, Truong orchestrated a series of limited thrusts to keep the NVA from massing for its final assault on Hue. A particularly effective tactic devised by the general and his American 1st Regional Assistance Command (RAC) advisers was the so-called *Loi Phong* offensive. Translating to "thunder hurricane," the stratagem involved concentrated and combined artillery, close air support, and naval gun bombardment against any NVA troop concentrations that could be found. The effects were devastating. *Loi*

Phong not only destroyed troop formations before they could fully organize but wreaked havoc on the North Vietnamese supply and communication lines necessary for a coherent assault on the city. To the tactic the U.S. Air Force added what Gen. John Vogt labeled "the most intensive in-country interdiction campaign of the war," as U.S. strike aircraft targeted nearly every bridge from the My Chanh River north to the DMZ. Mechanized columns of armor, artillery, troop transports, and supply convoys were left at the mercy of prowling fighter-bombers and B-52 Arc Light strikes as they desperately sought ways of crossing the region's numerous rivers and tributaries. But these efforts were not without cost. As NVA formations moved south, commanders steadily deployed ample anti-aircraft assets, including the SA-7 Grail surface-to-air missile launcher. Dubbed the 9K32 Strela-2 by its Soviet designers, the SAM was small and man- portable, making it far more mobile and difficult to locate and destroy versus its SA-2 big brother. North Vietnamese gunners would shoot down 10 U.S. aircraft over upper I Corps during April and May.[28]

Still, the situation in northern I Corps was improving. Seeking to build on the momentum, Truong launched a series of spoiling attacks to further flummox the enemy. U.S. helicopters airlifted two Marine battalions six miles south of Quang Tri City, where they ambushed elements of the NVA 66th Regiment on 13 May. A second air assault by 1st Division troops just days later managed to recapture the crucial FSB Bastogne. An ARVN ground element fought its way into the area a few days later, with the combined force effectively recapturing much of the territory lost during the previous month's fighting.[29]

Undaunted, the NVA finally launched its full-scale assault across the My Chanh River on 21 May. The communists managed to drive the Marine defenders back some three miles before South Vietnamese armored cavalry and intense airstrikes forced the invaders back across the river with heavy losses. A second NVA attack the following day yielded much the same result, as ARVN defenders again skillfully employed M72 LAW anti-tank rockets to chew up North Vietnamese armor. Yet another assault a few days later by troops of the 88th NVA Regiment, along with T-54/55 and PT-76 tanks, was again driven off with heavy losses as Marines in the northwest sector held firm. The attack marked the last significant attempt to take the city. Since moving into position near Hue in early May, the NVA had lost 3,000 killed and more than 60 tanks and armored vehicles destroyed. Still, although the immediate threat to Hue had passed, a strong NVA presence remained to the west and, most notably, to the north in the lost province of Quang Tri. After months of weathering the North Vietnamese onslaught, Truong now turned its eyes toward reclaiming South Vietnamese territory.[30]

The general devised a two-prong attack for his push north. But first, he had to ensure that NVA positions to Hue's west and south could not threaten either the city or his lines of supply and communication. The 1st ARVN Division was tasked with driving west to push the NVA farther away from the city. Meanwhile, the 2nd ARVN Division swept south to disrupt communist ambitions in Quang Tin and Quang Ngai provinces. On 16 June, the rebuilt 3rd Division, in its first duty after being mauled during the offensive's early days, deployed to the coastal city of Da Nang, 60 miles south of Hue. Despite Thieu's desire to have the "bad luck" division stricken from the rolls and renamed the 27th Division, Truong not only convinced his commander in chief to retain its unit designation but tasked the 3rd Division with protecting the crucial airbase and logistics facilities in the harbor city. With his southern flank shored up, Truong now turned his attention toward Quang Tri Province.[31]

The I Corps commander dubbed his general offensive "Lam

Son 72" and devised a two-pronged attack for the mission. The Airborne Division was tasked with the main effort. Its three brigades were to breach NVA lines across the My Chanh and then push north on the western side of Highway 1 toward the Thach Han River. The division was to take up positions on the southern outskirts of Quang Tri City. Meanwhile, the Marine Division, backed by the 1st Ranger Group and armor from the 7th Cavalry, was ordered to drive north up the coast along Route 555 and approach Quang Tri from the southeast. Thieu and his MACV advisers believed the plan too ambitious and wondered whether a series of limited spoiling attacks near Hue was a better option in the short term. Truong flew to Saigon on 26 June to personally sell the president on his plan. After some back and forth, Thieu finally gave his assent. The Second Battle of Quang Tri would commence on 28 June.[32]

The Airborne Division fought its way across the My Chanh with relative ease before encountering stiffening resistance as it approached the Thach Han River south of Quang Tri. Heavy close air support strikes softened NVA defenses, and the paratroopers' advance ground slowly closer to their objective. The Marines, however, were forced to fight a slogging advance up the coast against well-entrenched elements of the 304th NVA Division. Ferried by U.S. Navy transports, a two-battalion Marine amphibious assault behind enemy lines at Wunder Beach broke the logjam on 29 June. NVA units were forced to withdraw north rather than be cut off. While slower than expected, both lines of advance had already wrought destruction upon retreating NVA units. In less than a week, some 1,500 communist troops had been killed and nearly two dozen armored vehicles destroyed.[33]

By 7 July, elements of the 2nd Airborne Brigade had reached the outskirts of Quang Tri City. But there they encountered heavy resistance from a well-entrenched enemy supported by copious artillery and mortar fires. NVA forces were also firmly ensconced within the Old Citadel, whose walls were 13 feet high and 45 feet wide at their thickest point. The struggle for the Citadel would come to symbolize the grueling slog to retake Quang Tri over ensuing months. After two weeks of hard fighting, the paratroopers could neither dislodge the enemy from the fortress nor make very much progress in general. On 11 July, Truong ordered a Marine battalion to air assault northeast of the city to cut Route 560 north and interdict enemy resupply and communication lines. The amphibious assault ship USS *Tripoli* provided 34 helicopters for the lift. One CH-53D Sea Stallion was shot down by a heat-seeking SA-7 missile, making it the first U.S. Marine aircraft lost to the shoulder-fired SAM. U.S. Marines Staff Sgt. Jerry W. Hendrix and Cpl. Kenneth L. Crody were killed, along with 40 South Vietnamese Marines. Two other American helicopters were lost to ground fire and another 28 sustained damage. Once on the ground, however, South Vietnamese Marines killed 126 NVA troops and secured a foothold. After suffering some 150 casualties, the Marines were able to cut the road a few days later. On 22 July, a follow-on attack by the 147th Marine Brigade, along with a handful of tanks, dislodged the remaining NVA units and pushed them back toward the Cua Viet River. With the area northeast of the city now in South Vietnamese hands, the city was partially cut off. But rather than withdraw before the noose fully closed, Hanoi instead poured more men and materiel into Quang Tri. Elements of the newly arrived 325C and 312th NVA divisions even mounted a few modest counterattacks to drive back the airborne units to the west. The South Vietnamese effort soon faltered on all sides.[34]

Truong found himself between a rock and a hard place. The fight at Quang Tri had ballooned into an emotionally charged "cause célèbre" among Saigon's political and military leadership, and South Vietnamese public opinion. Withdrawal was unthinkable, and a stalemated siege was almost as unacceptable. Pressure to recapture the provincial capital was building to its bursting point. On the other side stood an entrenched and determined enemy who appeared hellbent on fighting to the last man. Truong decided to shake things up. The Airborne Division had spent the previous months as Saigon's de facto rapid reaction force, shuttling from one crisis to another to stave off defeat in the face of nearly overwhelming odds. Its ranks thinned and exhausted, Truong pulled the division off the line on 27 July. He then tasked Marine commander Brig. Gen. Bui Thi Lan with retaking the city. The Rangers were to defend the coast, while the newly redeployed paratroopers were ordered to protect the Marines' western flank and secure supply and communication lines along Highway 1.[35]

Complicating Truong's task was that up to six NVA divisions were now operating in Quang Tri and Thua Thien provinces, roughly twice what the general could call upon. Further, it became clear that Hanoi wanted to transform the fight to retake Quang Tri into a grueling, bloody, block-by-block quagmire for South Vietnamese forces. The NVA now had the 325C Division, along with the remnants of 308th and 320B divisions entrenched within the city. But if the NVA wanted to concentrate in Quang Tri, Truong would know just where to find them. As he would later write, this provided him with "the opportunity to accomplish my mission by employing the superior firepower of our American ally." Before long, U.S. strike aircraft were hitting NVA defenses throughout the city, while waves of B-52 Arc Light missions pummeled troop concentrations, and supply and communication lines north of Quang Tri. By the end of July, American aircraft would fly some 7,500 strike sorties and Arc Light missions in I Corps, most of them in and around Quang Tri City. Thousands of North Vietnamese troops perished or were wounded in what became the largest single-month number of strike missions of the Nguyen Hue campaign.[36]

But by August, the situation had remained largely unchanged. After a heavy artillery barrage, Truong sent in the 258th Marine Brigade to retake the Citadel. But the NVA were supported by their own artillery and mortars, and the attempt soon devolved into a brutal slog. Despite closing to within 200 yards of the old imperial fortress, the Marines were stopped cold and forced to dig in. On a brighter note, Truong's forces had gradually tightened the vice on Quang Tri, and the city was now surrounded on three sides. Perhaps most importantly, South Vietnamese troops were now firmly in control of the region northeast of the city, which blocked overland resupply from the recently captured port town of Dong Ha. Instead, the communists were forced to attempt resupply and reinforcement by ferrying men and materiel down the Thach Han River. These shipments quickly came under aerial attack, and stocks of food, medicine, ammunition, and troops replacements fell critically low.[37]

Even as the relentless air and artillery strikes ground the city to dust around them, the NVA remained determined to fight on, even launching a series of counterattacks over the first weeks of August. First, the communists tried to reopen their supply corridor to the northeast by pushing the 147th Marine Brigade off its blocking position on Route 560. A four-regiment sized assault force repeatedly hit the Marines but was repulsed with heavy losses each time. Then on 22 August, the NVA tried again, this time as troops within the Citadel attempted to clear out nearby Marine positions. After hurling some 3,000 artillery and mortar rounds at South Vietnamese defenders, communist infantry, along with a handful of tanks, boiled out of their defensive works and stormed the 258th Brigade position. But the Marines held firm, and the attackers were again driven back into the Citadel with heavy losses.[38]

By the second week of September, the drive to retake Quang Tri had entered its 10th week. Truong believed the time had come for decisive action. The enemy certainly appeared ripe for defeat. For more than two months, the communists had been pummeled night and day by air, artillery, and naval bombardment, their supply lines had been all but cut, and their casualties were mounting daily. Truong once again rebalanced his forces to bring maximum pressure to bear. First, he ordered Ranger units to replace the 147th Marine Brigade at the Route 560 blocking position. This freed up the brigade, which then shifted to assault positions on the city's north side. Next, the general ordered Airborne units to re-occupy former ARVN positions at La Vang southwest of the city. This would shore up the 258th Marines' flank as they pushed to retake the Old Citadel. Finally, with the U.S. Navy to ferry them, Truong ordered a diversionary amphibious landing of 400 Rangers at the mouth of the Cua Viet River 10 miles northeast of the city. The general hoped the feint would draw troops and artillery support away from Quang Tri's defenders. Truong had now provided Marine commander Brig. Gen. Lan two full brigades for an all-out assault on the city.[39]

After an intense artillery barrage, Lan launched his attack on 9 September. The 147th Marines on the northside drove toward the heart of the city, while the 258th Marine Brigade pushed east toward the Citadel. By 14 September, the brigade's 6th Battalion had blasted its way into the old fortress, gaining a tenuous foothold. Lan poured in more Marines to shore up the hard-won position. Meanwhile, one battalion of the 147th had successfully fought its way through the city and to the banks of the Thach Han River on the west side. Two other battalions drove south through the city and linked up with 258th Brigade positions outside the Citadel. The combined weight proved simply too much to bear for the beleaguered NVA defenders. By 15 September, the Citadel at last capitulated, with weary Marines hoisting the RVN flag the next morning. By late afternoon the following day, the last pockets of enemy resistance throughout the city had been eliminated. After more than four and half months under enemy control, the provincial capital of Quang Tri was finally back in South Vietnamese hands.[40]

As with An Loc and Kontum, President Thieu flew into the city to personally congratulate generals Truong and Lan. But the victory had been costly. The Marines had averaged 150 casualties per day over just the final 10 days of the assault. Since the beginning of the counteroffensive in June, the Marine Division had lost about a quarter of its fighting strength—some 5,000 casualties—3,658 of which were suffered during the effort to retake Quang Tri City. As for the NVA, the previous week's fight over the Citadel had alone cost 2,767 troops killed and another 43 captured. Meanwhile, more than two months of heavy combat, artillery, and airstrikes had pulverized the city to rubble, certainly resulting in thousands of civilian casualties.[41]

Over the next month, Truong set about solidifying RVN control over the rest of I Corps. The Airborne Division would eventually recapture fire support bases Anne and Barbara southwest of Quang Tri City, while the Marines solidified their lines south of the Cua Viet River. As with the aftermath of Nguyen Hue attacks in II and III Corps, however, South Vietnamese forces were simply too depleted to push remaining NVA forces back across the DMZ. Additionally, the river port town of Dong Ha on the Cua Viet to the north remained in North Vietnamese hands. The communists set about expanding its dock facilities, and within a year, more than 20 percent of the war materiel bound for the southern front would come through Dong Ha. The NVA had also gained control over the Laotian border area to the west and used the lull in fighting to repair and upgrade the Ho Chi Minh Trail in that sector. Over the previous two months, ARVN divisions to the south in Thua Thien, Quang Nam, Quang Tin, and Quang Ngai provinces had made slow but steady progress restoring governmental authority west toward the Laotian border. While ARVN commanders eventually succeeded in pushing the communists away from crucial population centers like Hue and Da Nang, and had secured the vital Highway 1 corridor, they were not able to fully expel North Vietnamese forces from RVN territory. Instead, the NVA, also battered and exhausted after months of heavy fighting, were content to retreat into remote and rugged sanctuaries like the Que Son Valley in western Quang Nam and lick their wounds. By the end of October, the military situation in I Corps had fully stabilized.[42]

The fight to reclaim Quang Tri marked the last gasp of Hanoi's Nguyen Hue campaign. In March, the communist high command had amassed the greatest conventional invasion force of the war, as some 225,000 troops and hundreds of tanks surged across South Vietnam's borders on three fronts. General Secretary Le Duan had banked on a stunning military victory that would annihilate South Vietnam's armed forces, expose Vietnamization as a failure, and push the United States out on terms dictated by Hanoi. What he got was the destruction of the North's best divisions, as some 100,000 troops fell dead and wounded, while hundreds of tanks, artillery, and antiaircraft weapons were left wrecked and smoldering on the battlefield. Le Duan and his advisers had clearly underestimated what airpower could do to such a mechanized force. Nguyen Hue's massed troop formations, and the extensive logistical and communication networks needed to support them, proved ripe targets for U.S. and RVNAF aircrews. It would take more than two years for the North's armed forces to recover.[43]

Nor had Hanoi's hoped-for popular uprising among the South Vietnamese citizenry come to pass. Just as it had in 1965 and 1968, Le Duan's strategy of the General Offensive-General Uprising had again failed in 1972. Far from greeting the NVA as "liberators," some 25,000

South Vietnamese had been killed and nearly 1 million made refugees as they fled invading communist forces.[44] Unfortunately for some of these civilians, however, Nguyen Hue had netted North Vietnam newfound control over about 10 percent of its southern enemy's territory. Though sparsely populated, most of the conquered land lay adjacent to the Ho Chi Minh Trail in the border regions of I, II, and III Corps. Its forces spent and reeling, North Vietnam contented itself with solidifying its position and upgrading and expanding its logistical networks in these areas.[45] The effort would pay dividends. The NVA's continued occupation of South Vietnamese territory would pose grave complications for Saigon in coming years.

South Vietnam's military had taken a beating, as well, and was in no position to immediately reclaim its lost territory. RVN figures listed 43,000 casualties and hundreds of armored vehicles and artillery destroyed or captured.[46] While a few South Vietnamese units broke and fled under the onslaught, most fought well, even courageously against a determined adversary that almost always enjoyed numerical and materiel superiority at the tactical level. U.S.- supplied M72 LAWs rockets proved especially useful in helping South Vietnamese ground troops stand in the face of NVA armor, exacting a dreadful toll in the process. Without doubt, the prodigious application of U.S. close air support, naval bombardment, and airlift capabilities—and the American FACs and Regional Assistance Command advisers who coordinated it all—played an invaluable role in helping stiffen South Vietnamese resolve. Of course, this continuing dependence on U.S. military assistance would portend ominous storm clouds for the years ahead. Still, though battered and bruised, South Vietnam's armed forces had taken the worst the communists could dish out and remained standing.[47]

Moreover, Nguyen Hue had wrought one dire consequence that

neither Le Duan nor the politburo had fully predicted. While an American response had certainly been anticipated—perhaps akin to the Rolling Thunder campaign of years past—few expected the ferocity and crippling effects of operations Linebacker and Pocket Money. Now, as Hanoi's offensive fizzled in the South, a plethora of newly freed U.S. air assets stood ready to hit North Vietnam even harder. And President Nixon had every intention of using them. Despite the resumption of formal peace talks in Paris on 13 July, he believed that one of his predecessor's key mistakes had been to extend frequent bombing pauses in hopes of advancing peace negotiations with Hanoi. But Johnson had erred, according to Nixon, in not seeing communist leaders for who they truly were— tough, "very pragmatic men" who respected strength and resolve. Rather than respond to Johnson's peace overtures, Hanoi had instead used the bombing respites to further bolster the country's air defenses, while continuing to infiltrate ever-growing numbers of men and materiel south. There would be no letup this time. Nixon determined that only the continued application of devastating airpower would convince Hanoi to negotiate in good faith. The president's decision surprised the communist high command, who were convinced the Americans would surely stop bombing once peace talks resumed. When they discovered this was not to be, as one observer later noted, it "impressed on North Vietnamese leaders the strength of Nixon's resolve to end the war only on acceptable terms." And so, as the summer heated up, so too did Operation Linebacker.[48]

7

NO LETUP

July - October

North of the 20th parallel, American airpower continued to pour it on. On 24 July, Air Force fighter-bombers destroyed supply complexes in Hanoi, the first strikes on the capital in a month.[1] Despite the service's focus on precision strikes in Hanoi and interdicting overland transport from China, the Air Force continued to fly mass strike missions using unguided bombs over the North Vietnamese heartland. However, as exceptionally poor weather blanketed the country's far north in August, it forced the service to fall back on its long-range all-weather navigation system (LORAN), which used multiple ground transmitters to communicate with strike aircraft crews and help triangulate their position and target location. Still, the early results were mixed, and the Air Force continued to lag behind the Navy in all-weather capability.[2] To keep up the pressure, sophisticated Navy A-6s and A-7s picked up the slack. The service, now with three carrier wings dedicated solely to pummeling the North Vietnamese heartland, would fly 5,000 sorties in August, the single-highest monthly total of the campaign.[3]

The Air Force attempted to catch up by improving its night-flying and all-weather capabilities. One hope rested on the F-111 Aardvark. Likes its animal namesake, the aircraft sported an exceptionally long nose. Housed within was the thing that made the

Aardvark special: its cutting-edge terrain-following radar (TFR) system. TFR enabled the bomber to skim the earth at just 100 feet while traveling more than 550 miles per hour—low enough to evade SAM radar, fast enough to prevent anti-aircraft gunners from drawing a bead. The anytime, any-weather F-111 boasted an exceptionally long range of 3,632 miles and could carry a payload four times that of the F-4 Phantom. As with most new technologies, however, the Aardvark went through early growing pains. In March 1968, six F-111s were deployed to the 474th TFW at Thailand's Takhli Airbase for a limited test in a combat environment. It did not go well. Three aircraft were quickly lost to mechanical issues, and further deployments were canceled.[4]

After further refinement stateside, 48 Aardvarks returned to Takhli in September 1972. The aircraft would go on to fly some 4,000 sorties over the duration of the war, usually against the most challenging and well-defended targets within the Air Force's area of responsibility. The F-111 would account for half the service's strike sorties in the high-threat arenas of Route Packs V and VI, tallying an impressive success rate for targets hit even in zero-visibility conditions. This was balanced against just six F-111s lost in combat, a remarkable mission-loss rate of just .02 percent. Because of the Aardvark's ability to strike targets accurately and without warning, the North Vietnamese would nickname it "whispering death." The F- 111 did not require fighter escorts, electronic warfare support, or inflight refueling, thus easing the burden of resource requirements and mission complexity for Air Force planners. In late September, the service continued to bolster its all-weather fleet by introducing the A-7D, an Air Force variant of the Navy's Corsair. Its own impressive 3,044-mile flight range further reduced the need for inflight refueling. By late fall, the Air Force was able to cut its tanker fleet by some 50 aircraft. Both bombers dramatically upgraded the service's all-weather strike capability, allowing it to shoulder more of the strike load over North Vietnam.[5]

Meanwhile, the Navy continued the life-and-death work of surface warfare in the Tonkin Gulf. On the morning of 26 June, the guided-missile destroyer USS *Benjamin Stoddert* was providing fire support for South Vietnamese troops fighting to reclaim Quang Tri Province when a shell misfired and became lodged in one of her 5- inch gun tubes. The ship's forward section was evacuated, and Lt. Cmdr. Michael J. Martin led five crewmembers into the turret to extract the live 70-pound projectile. During the nerve-racking procedure, the shell exploded without warning, blasting apart the turret and ripping a gaping hole in nearby compartments. Petty Officer 1st Class Robert T. Mills, 29, was killed instantly. Martin, who had been directing the work at the turret's open hatch, was blown overboard and also killed. Four other sailors were severely wounded. Senior Chief Petty Officer Gordon R. Uhler, 39, died of his wounds two days later. Seaman David N. Larson, just 23, survived for five weeks before he too succumbed on 8 August. The *Stoddert* underwent more than a month of extensive repairs at Subic Bay, Philippines before returning to the gunline at the end of July.[6]

The Navy's mining effort had proved instrumental in crippling North Vietnam's imports and limiting its warmaking capabilities. But it could also backfire. In the early afternoon of 17 July, the destroyer USS *Warrington* was running blockade and interdiction off the coast of North Vietnam when the ship was rocked on her port side by two underwater explosions some six seconds apart. The blasts severely damaged the ship's aft fire and engine rooms, as well as the main control room. Miraculously, no crewmembers were seriously injured, but the damage was so extensive that *Warrington* was forced to shut down propulsion and be taken under tow by another destroyer, the USS *Robison.* But the danger had not subsided. The ship began to flood, and crews struggled over the next several days just to keep her afloat. After being towed by three different ships, the *Warrington* eventually arrived at Subic Bay. Inspections in the following weeks revealed that the ship was too badly damaged to repair and was unfit for further service. On 30 September, the ship was decommissioned, making *Warrington* the largest U.S. vessel of the war to be permanently knocked out of commission. A subsequent investigation discovered MK 36 Destructor mine fragments within the ship's damaged hull. By 1 December, the Department of the Navy concluded that *Warrington* had struck two U.S. mines. Whether the mines had drifted from their original location, been improperly jettisoned by an American pilot, or whether the ship had steered into a designated mine jettison zone was never discovered. The *Warrington* would not be the last Navy vessel to suffer this unique form of "friendly fire." Just two weeks later, the destroyer USS *Hollister* struck another pair of Destructors. Suffering only minor damage, the ship was able to continue operations without delay.[7]

In mid-August, the JCS authorized Operation Lion's Den, a daring action that called for surface warfare ships to sail directly into Haiphong's heavily defended harbor and bombard nine high-value targets. Although Operation Pocket Money had effectively closed the port to shipping, Hanoi had since beefed up defense and other critical military assets in the vicinity, including a large ammunition dump, search and detection radar sites, coastal defense batteries, SAM sites, gun radars, and fire control direction centers. Under the overall command of Vice Adm. James Holloway, who in late May had assumed command of the Seventh Fleet, the flotilla was designated Task Unit 77.1.2 and was composed of the heavy cruiser *Newport News,* the guided-missile light cruiser *Providence,* the guided-missile destroyer *Robison*, and the destroyer USS *Rowan.* The group was divided into two equal elements. The *Newport News* and her 8-inch guns would pack the biggest punch and was tasked with entering the harbor approaches and bombarding the Haiphong emplacements. *Rowan's* primary responsibility was to screen the larger ship. But the destroyer's anti-submarine rocket launcher (ASROC) had been reconfigured to fire the anti-radiation Shrike, as well. The *Rowan* was to engage two targets in hopes of stimulating a North Vietnamese radar response. The destroyer could then target the site with its AGM-45s. It would be the first ship-to-shore use of the Shrike. Meanwhile, *Providence* and *Robison,* under Capt. John Renn, who had tactical command of the operation, were to hit targets just to the southwest, including the MiG airfield at Cat Bi.[8]

On 27 August, the task unit pulled into position and went to general quarters at 2200. Renn's element split off about 10 miles outside the harbor and busied itself with targets down the coast. *Newport News* and *Rowan* moved just beyond the mine nets laid in May and opened up a ferocious barrage. The harbor was well defended by a system of shore batteries composed of Soviet-built field artillery guns, including 152 mm howitzers. While the big cannons packed quite a punch, they were largely ineffective at hitting moving targets—especially at night. Should one of the American vessels become disabled in the channel, however, and the artillerymen could bring the stationary ship under murderous fire. Gunners let loose with their own voluminous wall of heavy shells, throwing up great geysers of salty spray all around the approaching U.S. ships. But the artillery lacked flash-less powder, making the big guns ideal targets for counterbattery fire. For more than half an hour, the two sides blazed away at one another, leaving targets throughout Haiphong ablaze, billowing pillars of fire and smoke curling into the night sky. The North Vietnamese had sent some 300 rounds in return, but despite a multitude of near misses, none found their mark.[9]

As Holloway's element turned to egress, the ship's combat

information center (CIC) suddenly lit up with four surface contacts speeding inbound. These were Soviet-built P-6 fast-attack torpedo boats, and they had been lying in wait to ambush the Americans as they exited the channel. The P-6 was armed with the Type 53 torpedo, its 650-pound warhead more than capable of sinking either ship. The *Newport News* banked starboard to unmask her guns and let loose with a ferocious broadside, blasting one of the boats to bits. But darkness and the blinding flashes of crisscrossing rounds prevented her from drawing a bead on the others. Holloway put out a call for illumination support. Within minutes two VA-93 A-7 Corsairs off the USS *Midway* arrived on station. The pair had been bound for an armed reconnaissance mission north of Hanoi when Holloway's call came in. As the million-candlelight flare drifted down, night became day in the darkened channel. The *Newport News, Rowan,* and the Corsairs opened up with enthusiasm. By 2243 it was all over, with at least one P-6 sunk, another damaged, and a third possibly sunk by one of the A-7s.[10]

Operation Lion's Den was deemed a success. All pre-planned targets had been hit, and at least three major secondary explosions were recorded. The flotilla's counterbattery fire had silenced several shore emplacements, as well. Though dozens of rounds had come close to hitting Holloway's element, both ships received only a light peppering with shrapnel. Renn's element also emerged unscathed. The task unit was disestablished the following day, and all ships returned to the gunline off Quang Tri Province. Tragically, the *Newport News'* good fortune would not hold. While providing gun support to South Vietnamese troops on 1 October, the ship would suffer a high-order in-bore explosion in one of the No. 2 turret's 8- inch guns. The explosion ignited some 750 pounds of powder in a horrific blast, killing 20 crewmembers and critically wounding another 36, many as a result of inhaling toxic gasses. It was a miracle that the ship was not completely destroyed. Incredibly, the *Newport News* would return to the gunline just three weeks later, but her No. 2 turret would remain sealed until the ship's decommissioning in June 1975. By war's end, 1,631 sailors would be killed in action, with another 4,178 wounded.[11]

As August turned to September, and the last vestiges of Hanoi's Nguyen Hue campaign fizzled out, more air assets became available for duty up North. Daily airstrikes rained throughout the country, as Air Force and Navy fighter-bombers hit industrial infrastructure, military installations, communication centers, transportation and logistics networks, and more. The North's infrastructure was indeed showing cracks. Earlier in July, Hanoi had dragooned all able-bodied civilians to work for a period of up to two months to help repair the damage. Two years' hard prison labor awaited any would-be shirkers.[12] In the North's southern panhandle, where the threat from SAMs was somewhat attenuated, more- vulnerable B-52s, A-4 Skyhawks, and F-8 Crusaders worked over enemy troop formations, and supply and communications lines from Vinh south to the DMZ. Nevertheless, North Vietnam managed to down another 19 U.S. aircraft for the month. In all, eleven airmen would perish that September, with another six captured.[13]

Still, Hanoi's situation continued to deteriorate. On 30 September, strikes on four airbases at Phuc Yen, Yen Bai, Vinh, and Quang Lang caught 15 MiGs on the ground, destroying five and damaging 10 others. This alone constituted 10 percent of Hanoi's remaining airpower.[14] Meanwhile, the Air Force's 432nd and 388th TFWs, marking a change in the service's fortunes, shot down 15 MiG 21s and 19s over September and October, losing just two Phantoms in return. The North had now lost some 62 of its best aircraft and pilots since the start of Linebacker. And the seaborne blockade and overland supply interdiction made them nearly impossible to replace. By September, imports had been cut by 50 percent.[15] Consequently, fewer and fewer MiGs now rose to challenge U.S. pilots over the heartland, even as stores of SA-2 missiles and even anti-aircraft ammunition dried up.[16] Nor could Hanoi import enough of the equipment and spare parts needed to repair the launchers and radar arrays that had survived. By fall 1972, the number of SAM launches and even the volume of anti-aircraft fire plummeted, with the number of U.S. aircraft shot down by SA-2s over September and October half that of Linebacker's first two months.[17]

Even as the renewed peace talks in Paris were starting show promise, air commanders offered no letup. On 1 October, 320 strike sorties hit targets throughout North Vietnam. Attack aircraft off the USS *Kitty Hawk* struck the military base at Phu Qui northwest of Vinh, blasting to rubble some 40 buildings, setting off secondary explosions and fires, and leaving the base in ruins. The *Kitty Hawk* continued to be a Linebacker workhorse. In the eight months since departing Naval Base San Diego, the carrier had spent 155 days on the line at Yankee Station, averaging 120 sorties per day. Incredibly, not even a mid-October race riot aboard the ship could slow the tempo, as carrier crews—some battered and bruised from the conflict—launched the sortie schedule right on time the next day.[18]

Also on 1 October, Air Force fighter-bombers hit Kep airbase and railroad yards near Hanoi, while B-52s worked over supply depots near Dong Hoi in the panhandle.[19] Air commanders followed up with another 300 strike sorties a few days later. On 11 October, Navy attack craft hit the railyard at Gia Lam in downtown Hanoi. During the raids, the nearby French diplomatic mission was heavily damaged, seriously wounding delegate general Pierre Susini, who would later die of his injuries. The nearby Algerian and Indian diplomatic missions were also impacted. The railyard had for years been off limits due to its proximity to civilian and diplomatic infrastructure. But two weeks prior, the JCS had finally added it to the target list. France and North Vietnam blamed the U.S., but American commanders insisted the damage resulted from a falling SA-2 missile, not U.S. ordnance. The Pentagon pledged to investigate. To ease tensions, Nixon imposed a temporary buffer around Hanoi the following day, ordering all future targets in the area first be cleared by the administration.[20] Nine days later, the Pentagon formally acknowledged that the bomb damage had been inadvertently caused by U.S. ordnance.[21] Still, air commanders kept up the heat. On 15 October, 350 fighter-bombers were hurled against targets ranging from the DMZ to the Chinese border, the second- heaviest day of bombing since the start of Linebacker. Air crews of the 388th TFW and 432nd TRW downed another three MiG-21s in the process.[22]

But rumors of peace were swirling in Paris. Since renewing the secret negotiations on 19 July, progress had been slowly building for national security advisor Henry Kissinger and North Vietnamese special envoy Le Duc Tho, with both sides giving ground on previously nonnegotiable issues. Now, as the air turned crisp, and the autumn leaves of Paris erupted in vibrant hues of red, orange, and gold, the momentum for settlement seemed to be gathering. Both sides, it appeared, had been mugged by reality.[23]

For Hanoi, the defeat of its Nguyen Hue campaign—and with it Le Duan's long-sought decisive military victory and general uprising—had convinced leadership that the bargaining table, not battlefield, would win the day. "We must concentrate our efforts…on our first objective," the general secretary declared, forcing "the Americans to withdraw."[24] Once the United States left, final victory over its "puppet" regime in the South was only a matter of time. Moreover, the ferocity of Nixon's air war had been very costly. It had not only stymied the nearly two-decade effort to conquer the South but had severely retarded the progress of socialism in the North. Since 1965, North Vietnam's population had grown by more than four million. But in the wake of

Operation Linebacker, the country's Gross National Product (GNP) had been knocked back to early 1960s levels. The longer the U.S. air campaign raged, the worse the North's economic and military prospects. Also lurking was the possibility that much-improved relations between the U.S. and Hanoi's Soviet and Chinese patrons might mean a reduction or even cut-off of foreign aid. From Hanoi's perspective, the sooner a settlement could be reached, the better.[25]

On the American side, the success of Linebacker and the defeat of the North's Easter Offensive were at first glance cause for optimism. But political and military imperatives had provoked a profound sense of urgency within the Nixon administration, as well. The president would almost undoubtedly be reelected in November. But it was Congress that worried him most. Democrats already enjoyed a commanding majority in both the House and Senate. And the incoming 93rd Congress on 3 January would likely further that dominance. If Nixon were not able to extricate the U.S. from Vietnam by then, years of congressional threats to cut off funding for the war—only intensified since the start of Operation Linebacker— were likely to become reality.[26]

As for the military situation, it was inextricably linked to the political. Even as he ramped up the air war to unprecedented levels, the president was steadily drawing down U.S. troop strength in Vietnam. By January 1973, just over 23,000 U.S. personnel would remain. Saigon would soon be on its own. If Nixon were to achieve "peace with honor," he would need an acceptable agreement, and soon. At minimum, Kissinger would have to obtain a deal that ensured the imminent release of the nearly 600 American POWs held throughout Indochina, while also providing Saigon breathing room before conflict with the North inevitably resumed. It was now up to Nixon's chief negotiator in Paris to achieve at the peace table what would not be won on the battlefield.[27]

8

PEACE AT HAND?

October – December

Although formal peace negotiations had resumed on 13 July, the real work of hammering out a settlement did not truly begin until Kissinger and Le Duc Tho renewed their "secret" meetings six days later. Gathering at a private home in the Paris suburb of Choisy-le-Roi, each began by reiterating longstanding positions, albeit with minor variations. These included ceasefire conditions, the exchange of POWs, the post-war political complexion of Saigon, and so on. But when it came to the presence of North Vietnamese forces in the South, it was Kissinger who blinked first. Without consulting RVN President Thieu, the Nixon administration was now offering to pull out all U.S. military assets from Indochina—without requiring a commensurate withdrawal of NVA troops from South Vietnam. Back on 8 May, Nixon had subtly excluded the long-standing U.S. demand from his televised address announcing the mining of North Vietnam's harbors and ports. "[North]Vietnamese troop withdrawal simply was not a significant issue," according to John Negroponte, who led the National Security Council's (NSC) Vietnam office in support of the Paris negotiations. "This was much more of a concern of the Saigon government than within our own delegation. And I do not believe that the Saigon government was sufficiently aware of the evolution of our own thinking in this regard." Still, the omission from Nixon's speech had not gone unnoticed by Thieu. The RVN leader had long suspected the administration would one day betray South Vietnam in the interest of ending America's role in the war.[1]

Thieu's fears appear well-founded. Nixon unquestionably desired an "honorable peace" meant to preserve U.S. credibility in the eyes of friend and foe alike. At minimum, this meant extricating the U.S. from Vietnam and securing the release of all American POWs, while providing the means for the RVN to survive once the U.S. withdrew. But how long the country could endure was an open question. According to transcripts of a 3 August 1972 Oval Office conversation between Nixon and Kissinger, it was crucial that any South Vietnamese collapse not come too closely on the heels of an American exit. This so-called "decent interval" would help ensure the United States would not bear responsibility for its erstwhile ally's defeat.[2]

Nixon believed that Linebacker and the blockade had "badly hurt" the North Vietnamese and that "the South Vietnamese are probably going to do fairly well"—at least in the short term. Ever mindful that the world was watching, however, Nixon asked, "But can we have a viable foreign policy if a year from now or two years from now, North Vietnam gobbles up South Vietnam? That's the real question." Kissinger believed the answer was "yes," provided the collapse "looks as if it's the result of South Vietnamese incompetence" and not wholesale abandonment by the United States. As to the ongoing negotiations, Kissinger warned, "If we sell out in such a way that, say, within a three-to-four month period, we have pushed President Thieu over the brink," and South Vietnam falls, the effects on American credibility globally would be very damaging. "So we've got to find some formula that holds the thing together a year or two, after which—after a year, Mr. President, Vietnam will be a backwater. If we settle it, say, this October, by January '74 no one will give a damn." Nixon agreed, likening the situation to the 1962 French exit from yet another of its former colonies. "Nobody gives a goddamn about what happened to Algeria after they got out." Still, Nixon was conflicted over leaving South Vietnam to such a fate. "But Vietnam, I must say . . . Jesus, they've fought so long, dying, and now… I don't know."[3]

Still, Kissinger's concession was stunning, especially in light of the allies' vastly improved military situation since North Vietnam's March invasion. Indeed, the resulting punishment had been the politburo's principal reason for resuming negotiations in the first place. But now, accepting the presence of NVA troops in the South had become official U.S. policy. The administration's about- face would haunt the peace talks—and Thieu's reaction to them—for the duration.[4]

For his part, Tho continued to insist that any communist forces in South Vietnam—including the Nguyen Hue campaign's nearly 200,000 troops, tanks, heavy artillery, and SA-2s—were merely "young southerners who had repatriated to the North in 1954 and young northerners who volunteered to go to the South to struggle." Hanoi had no control over these forces, claimed Tho, because they served in NLF units and were under PRG command.[5] Nevertheless, Tho made a show of protesting the proposal, complaining that it did not offer a specific timeline for American withdrawal and that it separated political and military issues. The latter criticism reflected a longstanding demand that any peace agreement be contingent upon Thieu's resignation and the dismantling of his "puppet" regime. Secretly, however, he was delighted. And so was the politburo, which eagerly authorized a follow-on session just 13 days later on 1 August. It was the shortest interval between meetings in nearly four years of secret negotiations. Both sides were cautiously optimistic about the prospects.[6]

As the sessions continued throughout August and September,

Hanoi became convinced it needed to reach a settlement with Washington before the November presidential election. Its reasons were manifold. First, it seemed assured that Nixon would be reelected. Hanoi had banked on getting a better deal from Nixon's Democratic opponent, Sen. George McGovern. McGovern had proudly positioned himself as the "peace at any cost candidate." He pledged that on Inauguration Day he would immediately stop all bombing of North Vietnam and withdraw all political and military support from South Vietnam—even if Hanoi had not released the American POWs. McGovern's declaration prompted columnist Joseph Kraft to question whether the candidate understood that to do so would mean accepting terms worse than Hanoi was actually offering.[7] But it seemed the chance of a McGovern presidency had all but passed. Now, Hanoi feared a reelected Nixon would have a fresh mandate to pursue an even-more aggressive military campaign. The failed Nguyen Hue Offensive had already cost the North dearly in men, military hardware, and materiel. Those losses, especially in the face of the U.S.'s ongoing blockade, would take years to replace.[8]

Second, the U.S-RVN pacification effort under the CORDS program had demonstrated real progress. Its efforts to win the "hearts and minds" of the rural population, along with the aggressive and effective Phoenix Program—which by mid-1972 had led to the death of some 26,000 communist insurgents, the capture of another 33,000, and the defection of about 22,000 to the South Vietnamese government—had become a serious cause for alarm in Hanoi. By March 1972, MACV's Hamlet Evaluation Survey concluded that some 97 percent of South Vietnam's rural population was living in "totally or relatively secure" villages. Hanoi had become so concerned over the program's results that it placed the disruption of the pacification effort high on the list of Nguyen Hue's objectives. Meanwhile, Vietnamization had swelled ARVN's ranks to more than 1 million strong, including

516,000 regulars. Coupled with a continuing influx of American military hardware, including rotary and fixed-wing aircraft, this represented an approximate doubling of South Vietnam's armed forces since 1966. Both factors suggested South Vietnam was growing stronger, not weaker, over time.[9]

Third, the politburo was increasingly concerned over the United States' growing détente and reproachment with its Soviet and Chinese benefactors. While the communist powers continued to support North Vietnam materially, Hanoi sensed cracks in the coalition. Were its patrons were more interested in courting improved relations with the Americans than in fostering the revolution in Vietnam? In a 19 August editorial, the VWP's official newspaper, *Nhan Dan*, complained that "the Communist powers had sacrificed proletarian internationalism to accommodate American imperialists," enhancing their own positions at the expense of the "heroic" struggles of the Vietnamese people. The longer the war went on, worried Hanoi, the more likely it was that Moscow and Peking would cut off support.[10]

Finally, the American air and blockade operations had taken a dreadful toll on North Vietnamese industry and infrastructure, imperiling not just the building of a socialist economy but the very revolution itself. Upon issuing new orders that the time had come to reemphasize propaganda and diplomatic efforts over the military struggle, Le Duan in late summer tacitly admitted that his audacious gambit had failed. In a letter to field commanders, the general secretary explained that, in addition to blockading North Vietnam's harbors and ports, "Nixon had resumed the war of destruction on a scale far surpassing that of the Johnson period. At this juncture, we must seize the opportunity and make resourceful use of the diplomatic weapon to fight and defeat the enemy. We must not only be resolute to win; we must know how to win." In other words, diplomacy had now become

North Vietnam's principal weapon for getting the Americans out of Indochina. As summer turned to fall, "the overriding aim was to get the United States out of Vietnam on the best basis possible, and keep her out," according to Truong Nhu Tang, the PRG cabinet minister for justice.[11]

By their 17th private session on 15 September, Kissinger and Tho had reached a watershed moment. After listening to Kissinger's newest proposal, Tho surprised the American diplomat by walking back one of Hanoi's longstanding inviolable demands—that Thieu resign, his government be dismantled, and South Vietnam's constitution be nullified. During their previous meeting, Tho had proposed creating a Provisional Government of National Concord (PGNC), a three-party coalition including representatives from the PRG, GVN, and a "neutralist" faction. The PGNC would govern South Vietnam until a permanent solution could be achieved. Though Tho still wanted a PGNC, albeit with diminished powers, he now proposed that the revolutionary government and Thieu's regime could continue in the short term. Each would be responsible for governing the areas it currently controlled and for working to implement coalition policies. For the first time in nearly four years of negotiations, Hanoi had dropped its demand for Thieu's resignation. Having witnessed the initial success of the Nguyen Hue campaign before U.S. airpower turned the tide, the politburo undoubtedly believed it could prevail once the Americans were out of the picture. Of course, to accept Tho's latest position would mean conceding North Vietnam's longstanding claim that the NLF and PRG were legitimate military and political entities in South Vietnam—a concession anathema to the Thieu government. But Nixon's national security advisor saw it as a sign that Hanoi was finally ready to settle.[12]

The American hastened to offer his own concession. He

dropped the administration's long-held position that it was possible to reach agreement on a ceasefire and the release of POWs before finalizing a comprehensive settlement. Hanoi had long worried that addressing political and military issues separately would leave the door open for U.S. reintervention in Vietnam once the North had surrendered its most important bargaining chip—the American POWs. Now, Kissinger had provided Hanoi the opportunity to shape the final agreement to prevent a continued U.S. troop presence in South Vietnam. "While we continue to object strongly to your holding our prisoners as hostages," Kissinger noted, "we are prepared to change our position. We now agree that implementation of the withdrawal and prisoner provisions would not begin until all negotiations are completed and an overall agreement is reached." Tho was elated. Another concrete step toward getting and keeping the Americans out of Vietnam had been secured.[13]

Following his latest concession, Kissinger noticed among the North Vietnamese delegates an "extreme eagerness to settle quickly." Tho told the envoy he believed a settlement could be achieved in a few weeks and asked to meet again as soon as possible. Finally, Tho eyed the American. "Do you really want to bring this to an end now?" he asked. Kissinger confirmed that he did. "Ok, should we do it by October 15?" Tho pressed. "That'd be fine," replied the American envoy. Both men appeared set on finalizing an agreement before the 7 November presidential election.[14]

Nixon was less convinced he needed an agreement that soon. To be sure, the president faced his own set of daunting imperatives to reach a settlement. On 3 January, the 93rd Congress would be sworn in and its Democratic majorities promised to be even more dominant—and hostile—than the outgoing legislature. In 1972 alone, legislators had introduced 35 resolutions to limit or end American involvement in

Vietnam. And then there was the growing Watergate scandal and its potential fallout. Both factors would significantly weaken his hand in the coming year. But the year was not yet over, and the president was not opposed to pouring on the pressure to force a more favorable settlement on Hanoi. It was true that Nguyen Hue—and his Linebacker response—had rallied Democratic opposition and the antiwar movement to McGovern. But Nixon believed the broader electorate was with him on Vietnam policy. Recent Harris polls seemed to bear that out, with voters favoring continued heavy bombing of North Vietnam by a margin of 55 percent to 32. A few days before Kissinger's 15 September meeting, Nixon met with Maj. Gen. Alexander Haig, 47, deputy assistant for national security and one of his most trusted aides. He directed Haig to convey his position to Kissinger prior to the session with Tho. "The president stated that the NSC (meaning Kissinger himself) does not seem to understand that the American people are no longer interested in a solution based on compromise, favor continued bombing and want to see the United States prevail after all these years," Haig cabled. "I emphasize to you his wish that the record you establish tomorrow in your discussions be a tough one which in a public sense would appeal to the Hawk and not the Dove."[15]

Nevertheless, Kissinger pushed for an agreement sooner rather than later. After meeting with Tho and Thuy for an 18th private session on 26 and 27 September—the first back-to-back sessions in the history of the talks—the national security advisor reported to Washington his confidence that an October settlement was in reach. "We can move to an announcement of the settlement sometime between the 20th and 30th," he declared. The agreement "would take effect with the ceasefire in place and a start of the release of prisoners in November."[16] Noteworthy, however, was the North Vietnamese proposal to form a new South Vietnamese national government—a Committee of National

Reconciliation— via an elected constituent assembly rather than a presidential election. Eligible candidates would include members of the PRG, providing Hanoi influence over South Vietnam's governance. This was tantamount to the dismantlement of Thieu's government, while potentially giving equal representation to a communist enemy that for nearly two decades had fought to conquer the South. While Nixon was not necessarily opposed to an early settlement, he was highly concerned that Thieu would reject the proposal out of hand.

The president dispatched Haig to sell the proposal to Saigon on 29 September. Predictably, Thieu objected to nearly every facet and was incredulous that Washington would want to settle on Hanoi's terms in the first place. If the Americans continued on this path, he said, "We shall be obliged to clarify and defend publicly our views on the subject."[17] Before departing Saigon, Haig vowed that "the United States would conclude no agreement with Hanoi without prior consultation." A few days later, the South Vietnamese president wrote to Nixon. The tentative agreement, declared Thieu, was surrender, plain and simple. "Everything will disappear," he wrote. "The president, the constitution, the general assembly, even the government itself."[18] Nixon responded with his own letter to steer Thieu toward acceptance. The message included both reassurances and at least one thinly veiled threat. "I give you my firm assurance that there will be no settlement arrived at, the provisions of which have not been discussed personally with you well beforehand." Nixon pledged to explore whatever "concrete security guarantees" could be wrung from Hanoi but also cautioned Thieu "to avoid the development of an atmosphere which could lead to events similar to those which we abhorred in 1963." Nixon, of course, was evoking the specter of South Vietnamese President Ngo Dinh Diem's overthrow and assassination in November 1963. The Kennedy administration had tacitly encouraged the coup, and Thieu himself had played a role as one of the military plotters.[19]

For its part, the VWP politburo was bent on ending the war as soon as an acceptable agreement was on the table. On 30 September, members gleaned from Tho and Thuy's reports that Washington did not appear to desire a pre-election agreement. This was worrisome because prolonging a settlement was not in North Vietnam's interest. "After the election," reported Tho and Thuy, "it will be difficult to compel the Americans to make additional concessions to those we can secure before the election." The envoys, therefore, recommended "forcing" Nixon into an early agreement. The politburo agreed. The long-term health of the revolution depended upon getting the Americans out as soon as possible. To that end, the CP. 50, the Hanoi advisory group charged with monitoring and analyzing the ongoing talks, prepared a draft proposal "mainly aimed at ending the American military involvement" and detailing minimum political requirements regarding the future of South Vietnam.[20]

Before the next meeting, the VWP cabled Tho and Thuy in Paris to ensure they understood Hanoi's position. At minimum, the men should steer Kissinger into an agreement that produced a ceasefire in place, halted the American air and naval campaign against the North, and removed all U.S. military forces from South Vietnam. The settlement should also acknowledge that there were two governments, two armies, and two zones of control in the South.[21] In other words, the PRG and NLF would be recognized as exercising exclusive political and military authority over the South Vietnamese territory they controlled. The politburo preferred that the war end before the U.S. election so as to "defeat Nixon's plan to extend the talks to win the election, continue Vietnamization of the war, and negotiate from a strong position."[22] It would be very difficult for the revolution to prevail, the politburo continued, without a change in "the balance of forces in the South," i.e., the withdrawal of all U.S. and other foreign military assets. If the Americans could be compelled into agreeing to withdraw, however, "we will have conditions to obtain these objectives later in the struggle with the Saigon clique and win bigger victories" down the road. The course set, Hanoi then cabled Washington to affirm its commitment to ending the war by the end of October. If the two sides did not reach an agreement, however, the war would continue "and the United States assumes responsibility."[23]

The 19th private session reconvened at Gif-sur-Yvette on 8 October. Kissinger began with a telling anecdote. On the way to the meeting, his car had passed a horseracing track. He had heard, said Kissinger, that one Paris racetrack featured a stand of trees that temporarily blocked patrons' view of the race. "You can't see them," he continued. "And I'm told that's where the jockeys decide who will win." Tho eyed the American. "Are we making a race to peace or to war?" he asked. Kissinger assured his counterpart. "To peace…and we are behind the trees!" Kissinger's meaning seemed clear. It was now up to he and Tho, hidden behind their own "stand of trees," to do what was necessary to finalize a settlement, even if that meant maneuvering around any reservations held by their respective governments.[24]

Kissinger was not far off, at least in his own case. Outside of a few close advisers, Nixon had by fall 1972 mostly cut himself off from subject specialists at Defense and State, and was increasingly in the dark diplomatically. Most of the people who knew what was really happening with the negotiations were in Paris. Haig was perhaps closest to the president in his belief that a better settlement could be had after the election. The two agreed that any preelection agreement that resulted in a public clash with Thieu was not desirable, especially if it could be interpreted as selling out to Hanoi. "[It] would hurt us more than it helps us," reasoned Nixon. But Haig, too, was in Paris, and Nixon was alone to wonder about what was happening. This left Kissinger with an unusually expansive role in shaping the talks as he saw fit. And the national security advisor knew that once a tacit agreement had been reached—and Hanoi were free to publicize it—anyone, including Nixon himself, would have a difficult time putting the genie back into its bottle.[25]

Later that afternoon, Tho handed Kissinger the new CP. 50 proposal, the first complete draft agreement ever submitted in nearly four years of talks: the "Agreement on Ending the War and Restoring Peace in Viet-Nam." It made official Tho's earlier offer to drop the longstanding demand that the Thieu government be dismantled and replaced with a coalition government featuring ample representation for the PRG. Instead, Hanoi proposed a National Council of National Reconciliation and Concord (NCNRC), what it described as an "administrative structure" designed to oversee the agreement's implementation following a ceasefire. The PRG and Saigon would continue to administer the territory they controlled. Kissinger was enthusiastic. This concession, he hoped, would help sell the agreement to Thieu, whom the envoy now saw as the chief obstacle. Kissinger pulled aside Winston Lord, his personal assistant. "We've done it," he said, shaking Lord's hand. That night, Kissinger cabled H.R. "Bob" Haldeman, Nixon's chief of staff. "Tell the president that there has been some definite progress at today's first session and that he can harbor some confidence the outcome will be positive." He also cabled the American ambassador in Saigon, Ellsworth Bunker. If the agreement went forward and resulted in a ceasefire in place, it was "essential that Thieu instruct his commanders to move promptly and seize the maximum amount of critical territory."[26]

Over the next two days, Kissinger and the North Vietnamese delegation hammered out the details. Nixon had little to go on save the nightly cables Kissinger sent through Haldeman. "The negotiations during this round have been so complex and sensitive that we have been unable to report their content in detail," Kissinger wrote. But he assured the president that, "We know what we are doing, and just as we have not let you down in the past, we will not do so now."[27] By 10 October, the two sides were comparing drafts and working urgently to move past any differences. Conflicts that could not immediately be overcome were simply tabled. Kissinger was again optimistic that a breakthrough was near and extended his trip into a fourth day. What followed was a grueling 16-hour session on 11 October that concluded with a provisional agreement featuring 18 articles divided into nine chapters. Kissinger departed Paris exultant.[28]

The envoy arrived in Washington just before 1800 and went immediately to brief Nixon. "Well, Mr. President," he exclaimed. "It looks like we've got three out of three! The opening to China, the beginning of détente with the Soviets…and peace in Vietnam."[29] Kissinger laid out the agreement's salient aspects. It featured a ceasefire in place with no expiration date, and required the U.S. to deactivate and remove all mines from North Vietnamese waterways. Sixty days following the ceasefire, all U.S. forces were to be withdrawn and a prisoner of war exchange was to take place. The North Vietnamese continued to insist that they had no control over any Americans held in Cambodia and Laos, so the fate of those POWs had been tabled. The question of some 35,000 detainees held by the Saigon government, dubbed "political prisoners" by Hanoi and the NLF, also went unresolved. Following its withdrawal, the United States, along with other foreign countries that had aided South Vietnam, was to cease support for the Saigon government and dismantle all military bases and facilities. While technically preserving the RVN government, at least in the short term, this concession effectively left Thieu to sink or swim on his own. The NLF was to accept no outside military assistance, either. The agreement made no mention of the 140,000 North Vietnamese troops in South Vietnam, effectively excluding them from the agreement's jurisdiction. In other words, if those forces violated the terms of the settlement, Washington would have no legal recourse under international law for punishing North Vietnam. This constituted a significant victory for Hanoi.[30]

Kissinger then turned to the political situation in postwar South Vietnam. While the North Vietnamese had dropped their demand that Thieu resign and his government be dissolved, the agreement put the PRG and Saigon government on equal footing. This was especially noteworthy since having the PRG recognized as a legitimate governing authority in the South had long been a goal of the insurgency. A National Council of National Reconciliation and Concord (NCNRC) consisting of three politically equal factions from the Thieu government, the PRG, and a purportedly neutralist element was to be created. The NCNRC would be tasked with "promoting the two South Vietnamese parties' implementation of the agreements, maintenance of the cease-fire, preservation of peace, [and] achievement of national reconciliation and concord ensuring democratic liberties." While not as powerful as the postwar coalition government the politburo had originally sought, Hanoi believed the NCNRC would be useful in strengthening communist influence over the political future of the South. As with the omission of NVA troops in South Vietnam, both of the above factors presented clear victories for Hanoi.[31]

Nevertheless, the council had the unenviable task of convincing the RVN and PRG to resolve their political differences within three months. At that time, "genuinely free and democratic general elections under international supervision" would be held. Particularly telling was the agreement's reference to just "three Indochinese states": Cambodia, Laos…and Vietnam. The language implied that the RVN and DRVN were not independent sovereign states. Since 1954, Washington and Saigon had maintained that, while the Vietnamese people embodied a single nation, North and South Vietnam were separate political entities with discrete governments, constitutions, and governing philosophies. As such, North Vietnam constituted a foreign power attempting to conquer its southern neighbor. Kissinger's accession to the new language effectively nullified the United States' rationale for nearly two decades of effort in Vietnam.[32]

Throughout Kissinger's presentation, an obviously anxious Nixon peppered the envoy with questions. With an agreement now in hand, the president seemed ready to end U.S. involvement as soon as possible. Kissinger concluded by declaring that he believed the settlement constituted "peace with honor" in Vietnam. The president eyed his chief negotiator. "Henry, let me tell you this," he said, "it has to be with honor. But also it has to be in terms of getting out. We cannot continue to have this cancer eating at us at home, eating at us abroad." But the president was especially worried over whether Thieu would go along. "Let me say, if these bastards [South Vietnamese leadership] turn on us, I—I am not beyond [unclear] them. I believe that's, that's what we're up against." Nixon then added, "I am not going to allow the United States to be destroyed in this thing." Haldeman, who was in attendance, summarized the group's main apprehension going forward. The "basic problem boils down to the question of whether Thieu can be sold on it." That unenviable task would fall to Kissinger. In the meantime, the national security advisor suggested a reduction in the bombing of North Vietnam as a show of goodwill. On 14 October, the president ordered bombing sorties over Route Packs V and VI cut to 200 per day. Two days later, he slashed the number to 150. Still, Nixon was determined to maintain some pressure. "I was not going to be taken in by the mere prospect of an agreement as Johnson had been in 1968," he later wrote.[33]

For their part, Tho and Thuy were delighted with the agreement.

In a cable to Hanoi immediately following the 11 October session, the envoys declared that the draft settlement had achieved five important objectives: it allowed PAVN forces to remain in the South, ended American involvement in the RVN, legitimized the existence of two governments, two military forces, and two zones of control in South Vietnam, obliged the U.S. to finance postwar reconstruction efforts, and acknowledged the rights of the South Vietnamese to decide their political future. The politburo shared their envoys' enthusiasm. The CP. 50 had crafted a document that guaranteed an advantageous military balance for the insurgency, restricted Saigon's "antirevolutionary" activities, and prevented the U.S. from militarily intervening to save its Southern "puppet." It seemed clear to Hanoi that, in the interest of extricating itself "honorably" from Vietnam, Nixon was willing to effectively abandon the Thieu regime. This comported perfectly with Hanoi's main objective in resuming peace negotiations. Get the Americans out once and for all. After which, the Saigon government's demise was just a matter of time.[34]

Kissinger returned to Paris on 17 October, meeting this time with Thuy to further refine the agreement. Questions remained over the status of American POWs in Laos and Cambodia, prisoners held by Saigon, and continued assistance to South Vietnam prior to the ceasefire, including the U.S. desire to replace worn and damaged RVN military equipment. The latter, codenamed Operation Enhance, had begun in May as a means of replacing on a one-for-one basis South Vietnamese military equipment lost during Nguyen Hue. On 20 October, Nixon would authorize Operation Enhance Plus to further bolster RVN military capabilities prior to American withdrawal. The program included the transfer of nearly 300 fighter- bombers, including more than 100 sophisticated F-5E Tigers, and the entire Southeast Asian fleet of A-1 Skyraiders over to the RVNAF. Additionally, 227 Huey helicopters, 32 C-130s, 200 armored vehicles, 2,000 trucks, guns and equipment for three artillery battalions, and large stocks of spare parts, ammunition, and POL supplies were to be included. Both sides hinted at flexibility on the issues, but nothing was finalized. Kissinger cabled Nixon saying he would try to gain some flexibility from Thieu on prisoner release in hopes of gaining Hanoi's acquiescence on weapon replacement.[35]

The envoy then flew to Saigon and presented himself at the Presidential Palace on the morning of 19 October. The American, accompanied by a retinue including Bunker, Army Chief of Staff Gen. Creighton Abrams, PACOM commander Adm. Noel Gayler, and other luminaries, began by presenting Thieu a letter from Nixon. After a preamble noting that Hanoi had at last dropped its demand for Thieu's resignation, Nixon came to the crux: "I believe we have no reasonable alternative but to accept this agreement…its implementation will leave you and your people with the ability to defend yourselves and decide the political destiny of South Vietnam." In keeping with Nixon's instructions, Kissinger then summarized the agreement at length, carefully emphasizing Hanoi's chief concession—that Thieu could remain in office—while minimizing and even omitting those elements Thieu would likely view as detrimental to the RVN's survival. Notably, the American did not mention that the settlement would allow 140,000 North Vietnamese troops to remain in South Vietnam. After his presentation, Kissinger handed the president an English-language version of the agreement. He withheld its Vietnamese counterpart, however, as well as the signing schedule he and Tho had already worked out.[36]

But Thieu knew something the Americans did not. Two days earlier his own Joint General Staff (JGS) had presented him with a document captured on 4 October from an NLF bunker in Quang Tri Province. Titled "Plan of General Uprising When a Political Solution is

Reached," the document not only summarized the agreement Kissinger had made behind Thieu's back but was a veritable how-to manual for manipulating the settlement in the field. The plan called for three phases. The first involved carefully studying the signed agreement to identify and exploit any loopholes prior to the ceasefire. Propaganda cadres were to requisition every available sewing machine and manufacture as many NLF flags as possible. Once hostilities ended, the flags were to be quickly raised over as many hamlets and villages possible to give the inspecting agency the impression that the PRG controlled the area. The second phase was to be initiated three days before the ceasefire, with local commanders staging coordinated attacks to gobble up additional territory. Meanwhile, political cadres would stir unrest and protests in urban areas to tarnish the Saigon government's image and shake public trust. The third phase would see the NLF consolidate its gains.[37]

It seemed that local Viet Cong commanders had known for weeks what Kissinger and Tho had in store for South Vietnam, including details of the settlement's provisions and even the signing schedule. Meanwhile, its president had been kept in the dark. "The Americans told me the negotiations were still ongoing and that nothing was fixed, but the other side already had all the information," Thieu later fumed. And now Kissinger was in Saigon presenting the agreement as a fait accompli. "I wanted to punch him in the face," Thieu said later.[38] The RVN president surely "was offended at being the last man consulted and then having no real voice because the matter was already decided."[39] Sensing resistance, Kissinger intimated that time was running out, raising the specter of a looming congressional cutoff of war funding. He urged Thieu to join the Nixon administration in presenting "this proposal as a victory [so that] we can prevail. If not, all we have striven for will be lost."[40] Thieu was buying none of it. His skepticism over the agreement would blossom into full-blown rage once he and his staff dug into the details.

The next day, Thieu kept Kissinger and his entourage waiting five hours before finally admitting them into the president's office. The Americans had delivered a copy of the Vietnamese-language version the previous evening, and the president and his advisers had spent hours preparing an initial list of the draft's most troubling provisions. In short, the agreement "was tantamount to surrender," said Hoang Duc Nha, Thieu's cousin and trusted adviser. "What are these 'three Indochinese states' that are referred to?" Thieu demanded. The idea that there was only "one Vietnam" completely undermined the principle of RVN sovereignty. South Vietnam, which had not signed the 1954 Geneva Accords, nevertheless had long interpreted the agreement's DMZ at the 17th parallel as a firm political boundary separating two distinct countries.[41]

Kissinger, who had relied on North Vietnamese translators in Paris, could not read the Vietnamese-language version. Nevertheless, he reassured Thieu that the reference "must be a typographical error"—although the phrase "three Indochinese states" appeared three times in the English text. The South Vietnamese then turned to the NCNRC, which the English version had depicted as a "committee" or "council." Yet the Vietnamese text described it as a "governing structure." How was this very much different than the coalition government the Americans and South Vietnamese had long rejected? Also of note, Thieu continued, was that the Vietnamese version disparaged the Americans as "pirates" and the South Vietnamese as "vassals." Yet another point of contention was the equal footing granted to the PRG in postwar South Vietnam. This placed the NLF's recently created political entity on par with the RVN government, effectively legitimizing an enemy organization that had long sought its overthrow. Above all, however, Thieu and his advisers vehemently opposed allowing North Vietnamese troops to remain in their country. This represented a clear and present danger to the RVN's very survival, said Thieu. He then related a story to Kissinger. A man catches a thief in his bedroom and calls police to make him leave. After failing to convince the thief, the police chief gives up and holsters his weapon. "He's not such a bad guy," the chief tells the homeowner. "Why don't you try to learn to live with him? After a while he may get homesick and go back to his own family…or, he might try to rape your wife."[42]

Thieu's apprehension only grew later that day when he saw a *Newsweek* interview with North Vietnamese Prime Minister Pham Van Dong. Dong had contacted American journalist Arnaud de Borchgrave and requested the interview, a violation of the longstanding agreement between Hanoi and Washington to keep details of the covert peace talks secret. Dong used the opportunity to both pressure Nixon into finalizing the agreement and to drive a wedge between Washinton and Saigon. And he was not above lying to do so. The prime minister said that once the Americans withdrew there would indeed be two governments and two armies in the South. Hinting that the NCNRC might be a coalition government after all, he mentioned a "three sided coalition of transition" to "promote democracy and speed national concord in the south." Dong confirmed that his government desired Thieu's ouster and that for the Vietnamese "reunification is in our blood, in our hearts." He also lied when he proclaimed that all detainees—American POWs and Saigon's prisoners, alike—were to be released at the same time. Kissinger, of course, had been clear that only Saigon could negotiate the political detainees' release. All of this confirmed the South Vietnamese president's suspicions that not only would the agreement be catastrophic for his country but that Washington and Hanoi had been conspiring behind his back.[43]

Despite Dong's maneuver, Hanoi and Washington were inching

closer toward final agreement. In a cable to Nixon, the North Vietnamese reversed their position on three of Kissinger's 17 October proposals that the American envoy had considered major sticking points. Hanoi now agreed to a piece-for-piece replacement of worn or damaged RVN military hardware under international supervision. The North also accepted Kissinger's phrasing on the POW issue, which had asserted that prisoners held throughout Indochina—not just North Vietnam—must be covered in any peace agreement. Hanoi had long claimed that it was not accountable for prisoners held by communist insurgents in Cambodia and Laos. The politburo's about-face now implied that it would take responsibility for seeing to the release of those detained by its allies. Finally, Hanoi dropped its demand that Thieu free political prisoners as part of the initial settlement. Instead, the North accepted Kissinger's suggestion that prisoner status could be negotiated between the PRG and RVN within three months of a ceasefire. Having swept aside what it considered to be the last impediments to peace, Hanoi deemed the negotiations complete and asked the U.S. president not to submit further changes. The message reiterated the Kissinger-Tho October timeline for signing the final agreement, and declared that, "should the U.S. side continue to seek pretexts to delay the schedule," negotiations would break down, the war would continue, and the "U.S. side should bear full responsibility." A follow-on message threatened to go public with the settlement's details if Nixon did not sign by the following day.[44]

Nixon replied the next day and thanked Dong for making the concessions. Contradictorily, the president stated that "the text of the agreement can now be considered complete," but then went on the make additional demands. Nixon agreed to North Vietnam's declaration that the two sides should work toward a ceasefire in Laos, end all military activity there, withdraw all forces, and not interfere with the peace process. But the president wanted the same commitments on Cambodia, as well. Most importantly, Nixon demanded written assurances that Hanoi would do whatever was necessary to secure the release of POWs in Laos and Cambodia according to the same timeline laid out for Vietnam. A nonnegotiable pillar of the president's "honorable peace" was the release of all American POWs in Indochina. Without ironclad guarantees, the United States would never sign an agreement to end the war. Once those additional conditions had been met, according to Nixon, "the agreement can be considered complete."[45]

Meanwhile, Kissinger had his hands full with Thieu. Together with Bunker, the American laid out Hanoi's latest concessions and poured on the pressure to gain Thieu's acceptance. But the RVN president was unmoved. The agreement was nothing short of capitulation, he said. The United States was not only abandoning South Vietnam but was reneging on its commitment to prevent the spread of communism throughout Indochina. Thieu's Foreign Minister Tran Van Lam presented the American with a list of 23 changes the South Vietnamese wanted to see in a revised agreement. Among the most prominent were Saigon's demand that the phrase "administrative structure" regarding the NCNRC be stricken from the text and that language be inserted sanctioning the 17th parallel as a legitimate political boundary. This, it was hoped, would affirm RVN sovereignty and render any northern attacks below the DMZ illegal. A frustrated Kissinger promised to press for the changes.[46]

Doubt was beginning to form back in Washington. Conferring with Nixon, both Haig and the recently retired Westmoreland, who had returned to lend his counsel, expressed grave concerns over whether Thieu would—or even could—accept the proposal. Westmoreland was especially concerned that the agreement was "not adequate to the realities of the situation" and worried that if the Americans imposed this settlement on Thieu— effectively making him the "puppet" Hanoi had always claimed him to be— it "could prove fatal to Thieu's own political base." Instead, said Westmoreland, the U.S. should "work patiently with Thieu and recognize the difficulty that relinquishment of his territory would pose." According to Haig, Nixon had "no intention of being stampeded." The president cabled Kissinger in Saigon. "We must have Thieu as a willing partner in making any agreement," the president wrote. "It cannot be a shotgun marriage."[47]

Even Kissinger had to admit the situation looked perilous from Thieu's perspective. At the end of another long day of fruitless negotiation, the envoy got word to Washington. The South Vietnamese "undoubtedly feel they need more time…they always feel that way," he wrote. "They know what they have to do but it is very painful. And they are probably even right." After long depending on U.S. help to stand against Hanoi, he continued, "their nightmare was not this or that clause but the fear of being left alone. They could not believe Hanoi would abandon its implacable quest for domination of Indochina. They were panicky…and they were not wrong."[48]

That perspective did not last long. Pressure built as domestic and international press organizations caught wind of the rumored peace. By 22 October, dozens of print and broadcast outlets around the world were running stories heralding an impending peace agreement. Even Wall Street responded as the Dow Jones average leapt more than 10 points. The genie was indeed out of the bottle. Meanwhile, Nixon received a message from Pham Van Dong criticizing him for missing the signing deadline. The politburo, said the prime minister, had met every one of the president's most recent demands. Dong added that the Pathet Lao, the Laotian communist insurgency and longtime ally of North Vietnam, had affirmed its openness to a cease fire and the release all American POWs by 30 December. The North would also work to broker a ceasefire in Cambodia. Not mentioned was the fact that tensions with the communist Khmer Rouge had led to a marked decline in Hanoi's influence with the Cambodian insurgents. Nevertheless, Dong said, Hanoi had gone as far as it was willing to go. It was time to finalize the agreement and bring the war to a close.[49]

Kissinger cabled the president to reinforce the urgency of the situation. "We have obtained concessions that nobody thought were possible last month…or last week," he wrote. Referencing the North Vietnamese, he warned that, "we are dealing with fanatics who have been fighting for 25 years…we cannot be sure how long they will be willing to settle on the terms now within our grasp." Similarly refocused, Nixon ordered his envoy to convey a message to Thieu. "May I read you the telegram from President Nixon?" Kissinger asked. Thieu nodded. Kissinger spoke the words of his president: "Were you to find the agreement to be unacceptable…and the other side were to reveal the extraordinary limits to which it has gone in meeting demands…it is my judgement that your decision would have the most serious effects upon my ability to continue to provide support for you and for the government of South Vietnam." Nixon's implication was clear. *Get on board or you're on your own.* Thieu did not flinch. In its present form, the agreement jeopardized the very existence of South Vietnam. "I do not appreciate the fact that your people are going around town telling everybody that I signed," he said. "I have not signed anything. I do not object to peace, but I have not gotten satisfactory answers…and I will not sign."[50]

Kissinger exploded. "We have fought for four years…mortgaged our whole foreign policy for the defense of one country," he exclaimed. "You're the last obstacle to peace. If you do not sign, we're going to go on our own." Was this Thieu's final answer? "Yes, that is my final position," Thieu replied. "I will not sign.

Please go back to Washington and tell Mr. Nixon I need answers." Kissinger did not wait, cabling Nixon that night. "Thieu has just rejected the entire plan or any modification of it," he wrote. "It is hard to exaggerate the toughness of Thieu's position. His demands verge on insanity." Haig had warned earlier that the last thing the administration needed was for Hanoi to agree to all U.S. demands while Thieu remained obstinate. It seemed that dark portent had become reality.[51]

From Thieu's perspective, the refusal was the only rational and dignified course of action. The settlement failed to establish the DMZ as a legitimate political boundary, placed too much power into the hands of the NCNRC, a council that appeared a coalition government in all but name, and allowed some 140,000 NVA troops to remain in his country. Moreover, the PRG was to be afforded political legitimacy and ceded South Vietnamese territory, placing it on an equal footing with the GVN. This was nothing short of a travesty, according to Thieu, and in no way reflected the actual balance of political forces in the South. All of which, he believed, represented a clear and present danger to RVN survival. But the president also felt betrayed, saying that "those whom I regard as friends have failed me." His American allies had pledged that nothing would be settled without his full consultation and approval, but then turned around and negotiated an all-but-finalized settlement behind his back. And now Kissinger—a man he both disliked and distrusted—was insisting he must either submit or be abandoned by the United States.[52]

Kissinger was beside himself. In a pique, he recommended that the United States go it alone. Nixon should end all bombing of the North and sign a bilateral agreement with Hanoi. Bunker would have none of it. To do so—without at least wringing further concessions from Hanoi—would severely damage American credibility with other Southeast Asian allies who looked to the U.S. for support against communist expansion. Haig, too, sympathized with Thieu's plight. He cabled Kissinger in Saigon. "You should not underrate the substantive justification for Thieu's intransigence," he wrote. "He…is being asked to relinquish sovereignty over a large and indescript portion of South Vietnamese territory," he wrote. "He has never agreed to such a concession and given his paranoia about what has brought us to this point, it is understandable that he would now accept an open break." While Kissinger quickly realized that a bilateral agreement with Hanoi would harm U.S. interests, his resentment toward Thieu only deepened. His deal was dead, and the envoy's working relationship with Thieu was in shambles. It would never recover.[53]

If Kissinger were in search of where to lay the blame, he might have tried looking in the mirror. The October negotiations had indeed produced an agreement that satisfied Washington and Hanoi. For the U.S., "satisfaction" meant gaining the release of its POWs and withdrawing from a Vietnam that would not immediately collapse—the so-called "decent interval." On the DRV side, it meant getting the United States out of Vietnam once and for all, while providing the military and political conditions for an eventual victory over the South. But it did not satisfy the needs of the country over which nearly two decades of war had been fought. In his eagerness to secure a historic peace agreement—the most consequential act of his long career, he later claimed—Kissinger pushed forward with a set of U.S. concessions that the South Vietnamese president would surely reject. The most troublesome, according to Thieu, granted legitimacy to the PRG and put it on an equal footing with Saigon, imposed a "coalition government" (the NCNRC) on his country, and called for a ceasefire-in-place that would cede large swaths of RVN territory to the communists. The worst by far, however, was that some 140,000 NVA troops—poised like a dagger at Saigon's throat—would remain in South Vietnam. Moreover, Kissinger had agreed to these concessions while repeatedly breaking his country's pledge that no settlement would be finalized without first fully consulting Thieu and gaining his assent. Instead, the draft agreement had been "negotiated over the head of the GVN," with its president kept ignorant of the proceedings. With Nixon's blessing, Kissinger then traveled to Saigon to foist the settlement on a man who had no hand in its negotiation. According to Thieu, to accept such an agreement would not only threaten his country's survival but would undermine his credibility as a national leader. Hanoi, he would later say, "had always accused us of being America's puppet. Now Kissinger was treating us like one. There was no effort to treat us as an equal."[54]

By this point Nixon, already open to postponing an agreement until after the election, now seemed resolved to that course. He did, however, wish to keep the North Vietnamese talking so that "we can maintain [an] aura of progress through November seventh." Besides, if the North Vietnamese had agreed to U.S. demands up to this point, there was at least an even chance they might give more ground once he had been safely reelected. The president broke the news to Pham Van Dong. After reaffirming his commitment to the substance of the agreement, he came to the crux: "Unfortunately, the difficulties in Saigon have proved somewhat more complex than originally anticipated. Some of them concern matters which the U.S. side is honor-bound to put before the DRV side." He asked for a continuance of the talks and urged "restraint" in the short term, the latter an admonition to keep details of the negotiations confidential. Nixon also implied that Dong's *Newsweek* interview was at least partly to blame for the troubles with Saigon.[55]

On 23 October, Nixon kept the prospects for peace alive by halting all bombing above the 20th parallel. To ensure Hanoi did not use the pause to resume unfettered infiltration into the South, air attacks in the Southern Panhandle and along the Ho Chi Minh Trail would continue. Since 10 May, U.S. aircraft had flown nearly 42,000 sorties against North Vietnamese military assets and infrastructure, unloading more than 155,000 tons of bombs in the process. The war's most intense air campaign, along with Operation Pocket Money, had largely accomplished the three main goals set by the administration: cripple North Vietnam's warmaking capacity by curtailing foreign imports and pounding its military, economic, and industrial infrastructure; strangle Hanoi's Nguyen Hue Offensive by cutting the southward flow of men and materiel; and convince Hanoi to resume serious peace negotiations. That success had come at a cost, however. Some 99 U.S. aircraft had been shot down over the five and a half months of Linebacker, including 45 Air Force, 44 Navy, and another 10 Marine aircraft. More than 60 percent of Air Force losses resulted from MiG engagements, while the Navy/Marines lost roughly the same percentage to ground fire. SA-2 missiles would claim 24 U.S. aircraft. Finally, scores of U.S. airmen had been killed during the air campaign and another 70 taken prisoner. While mining and blockade efforts under Pocket Money would continue, Operation Linebacker was officially over.[56]

§

Both Vietnamese nations took the secret Paris negotiations public the very next day. Addressing a national radio and television audience, Thieu spoke for two hours to both reassure his nation and to stake out his position on the negotiations. He affirmed that the alliance between the United States and South Vietnam remained strong and that leaders from both countries would ensure the RVN's survival. Without revealing the proposal's specifics, he nevertheless pledged to reject any peace agreement that imposed a coalition government on the RVN or that permitted NVA troops to remain in the South. In an obvious end-around of the Kissinger-Tho negotiations, Thieu proposed that South and North Vietnam negotiate directly to broker their own agreement. The president's address found a receptive audience. By the end of October, his popularity surged in South Vietnam. According to one monitoring agency, the International Commission for Supervision and Control (ICSC), Thieu enjoyed his "strongest support ever" in both rural and urban areas.[57]

A few hours later, Radio Hanoi went public with its own messaging. The North Vietnamese released a chronology of the talks, certain key terms of the draft settlement, and the agreed-upon signing timeline. Hanoi was careful, however, to divulge only the elements that made it appear as if Hanoi had met all U.S. and RVN demands. The communists accused the U.S. of backing out of its pledge to sign and implement the settlement and demanded it be finalized by 31 October as agreed. Xuan Thuy held his own news conference in Paris, repeating the accusation that the U.S. had gone back on its word and insisting that Nixon's request for more talks created an "extremely dangerous situation."[58]

Both events caused concern in the White House. Nixon decided it was time to break his so-called "Kissinger Rule," the unspoken policy of keeping his national security advisor—and his German accent—off American TV sets. On the morning of 26 October, Kissinger held his first televised news conference to explain the administration's position. "We have heard from both Vietnams, and it is obvious that a war that has been raging for 10 years is drawing to a conclusion," he intoned. "We believe that peace is at hand… that an agreement is within sight based on the May 8 proposals of the president and some adaptations of our January 25 proposal which is just to all parties."[59] Reporters asked why the deal had not been made in 1969.

"The other side consistently refused to discuss the separation of the political and military issue," Kissinger replied. "It always insisted that it had to settle the political issues with us, and that we had to predetermine the future of South Vietnam in a negotiation with North Vietnam." Kissinger's "peace is at hand" declaration rocketed throughout newsrooms across the country and world. Influential *New York Times* journalist James Reston, one of Nixon's most vocal critics, nevertheless lavished his chief negotiator with praise in a column titled "The End of the Tunnel." *Newsweek*'s cover announced "Good-Bye Vietnam," complete with a feature story explaining "How Kissinger Did It." Given the rampant speculation in news reports over the previous weeks, it was probably inevitable that reporters—and the world— would believe an agreement was imminent. Did Kissinger truly believe his claim, did he simply misspeak, or did Nixon's envoy intentionally mislead?[60]

In a 27 October classified memo, Kissinger implied that an agreement could be finalized "in matter of weeks," provided both sides continued to exercise "good will" in their negotiations. The next day, Kissinger's deputy, William H. Sullivan, intimated to Canadian diplomats that the "earliest date for [a] cease-fire could come two or three weeks hence."[61] Whether the above represent post facto damage control is unknown. Kissinger later wrote that, while he regretted the phrasing, it represented what he believed to be true at the time. "It was a pithy message—too optimistic, as it turned out—to the parties of our determination to persevere; a signal to Hanoi that we were not reneging and to Saigon that we would not be derailed. And despite all the opprobrium heaped on it later, the statement was essentially true—though clearly if I had to do it over I would choose a less dramatic phrase." Nixon, he wrote, did not know that his national security advisor would use that expression. Caveats aside, Thieu's vehement rejection of the draft settlement made Kissinger's optimism for a finalized agreement "within weeks" appear wholly unrealistic.[62] Nixon was far less sanguine in his assessment. "When Ziegler told me that the news lead from Kissinger's briefing was 'Peace is at hand,' I knew immediately that our bargaining position with the North Vietnamese would be seriously eroded and our problem of bringing Thieu and the South Vietnamese along would be made even more difficult," he later wrote.[63] Haig shared his president's appraisal. "It is hardly possible to imagine a phrase, so redolent of Neville Chamberlain and the effete 1930s cult of appeasement, more likely to embarrass Nixon as President and presidential candidate, inflame Thieu's anxieties, or weaken our leverage in Hanoi," Haig recalled in his 1992 memoir. "The President regarded Kissinger's gaffe as a disaster."[64]

None of the above seemed to impact Nixon's bid for a second term. On 7 November, he was reelected in a landslide, carrying 49 states (including McGovern's own South Dakota) on his way toward winning 60.7 percent of the national popular vote. The president became the first of his party to win a majority of Catholic voters, picked up 50 percent of the youth vote, and even captured more than a third of Democrats. With the greatest margin of victory in American political history, Nixon's "great silent majority" had unequivocally rejected the "peace at any cost" candidate. But that landslide did not translate to the rest of his party. Nixon's fears for the composition of the new Congress had been well-founded. Republicans lost two seats in the Senate, tightening the Democrats' grip on the chamber 57 to 43. His party did manage to pick up 12 additional House seats, but Democrats retained a dominant 243-192 edge. Despite achieving what NBC *Nightly News* anchor John Chancellor called "the most spectacular landslide election in the history of United States politics" Nixon knew any "mandate" he had received would begin to crumble once the 93rd Congress was sworn in on 3 January. Meanwhile, a few weeks earlier, *Washington Post* reporters Bob Woodward and Carl Bernstein had published their first major frontpage piece on the White House's burgeoning Watergate scandal. As other major news outlets joined the fray, attention to the story was only gaining steam. Time was running out.[65]

§

The first weeks of November were marked by accusations and recriminations on all sides. Hanoi accused the United States of breaking its word by failing to sign the 11 October draft agreement and asking for further negotiations. Meanwhile, Thieu had begun his own information campaign to both undermine the Paris negotiations and to shore up support within his own government. He also sent several personal letters to Nixon pleading for the U.S. president to walk back Kissinger's many concessions—especially the acquiescence to some 140,000 NVA troops remaining in South Vietnam. Those forces not only posed a practical threat to the RVN's survival, but permitting them to remain undermined the very notion of South Vietnam as a sovereign state with legitimate political boundaries. Nixon reciprocated with his own slew of correspondence, urging Thieu to get on board with the agreement. For as much as Nixon wanted to end American involvement in Vietnam, "peace with honor" required Thieu be a willing partner. He wrote that the allies must present a united front to both Hanoi and to the U.S. Congress. If Thieu demonstrated a good-faith willingness to make peace, it would be that much easier to convince Congress to support South Vietnam in the face of any "massive violation" by Hanoi. After all, as Kissinger would later say, the terms of any agreement ultimately meant very little to Hanoi. What the politburo cared most about was "whether the B-52s may come again." Unity between the allies would convince Hanoi to honor the agreement.[66]

The private negotiations resumed in Paris on 20 November but made very little headway. Saigon's list of required changes had grown from 23 to 69 items, the most significant of which continued to demand that North Vietnamese troops withdraw from the South. Over coming days, Kissinger attempted to walk the fine line between clawing back just enough of his earlier concessions to gain Thieu's blessing without blowing up the deal entirely. But his entreaties fell on deaf ears. Hanoi had instructed Tho and Thuy to refuse any of Kissinger's overtures, instead ordering them to continue calling for the U.S. to honor the terms of the October draft agreement. Instead, the North Vietnamese envoys fell back on the well-worn tactic of using the talks to inveigh against the U.S. and its "puppets" in the South. When the DRV delegates did engage on a substantive level, it was to walk back many of their own concessions from the October draft. Chief among them was to re-link the release of American POWs to that of political prisoners in the South. The envoys also revived Hanoi's longstanding demand that Thieu resign following the ceasefire. In his report to Nixon, Kissinger relayed that the North Vietnamese "demonstrated absolutely no substantive give and in fact drastically hardened their position on the political conditions…[and] in several important areas they returned to former (pre-October 20) negotiating positions."[67]

Meanwhile, to maintain leverage at the bargaining table, the politburo ordered its forces in the South to reverse course once again. Worried over the seeming mandate of Nixon's landslide reelection and fearing it would stiffen the U.S. president's resolve, the politburo looked again to the "strategic balance" of forces in the South.

Commanders were to once more put political and propaganda efforts on the back burner and return to the "military mode of struggle." Over ensuing weeks, attacks around the Thach Han River, Dong Ha, Ai Tu, and the Cua Viet region saw PAVN forces seizing ground.

Additional forays in the Central Highlands netted control over some 95 hamlets and about 13,000 villagers. Attacks near My Tho, Ben Tre, and Bac Lien further increased North Vietnamese gains, adding another 1,000 hamlets and some 100,000 people to its areas of control.[68]

After six days of fruitless negotiations, Kissinger on 25 November proposed a 10-day break. Tho agreed and used the opportunity on 28 November to send Hanoi a summary of the recent negotiations. In general, Kissinger's proposed changes were wholly unacceptable because they upended the spirit and substance of the October agreement, reversing "all important issues." Among the Americans' "absurd" demands were the "withdrawal of Northern forces, the reduction and demobilization of soldiers, the elimination of each side's established zones of control…the elimination of the three-component [the administration—the NCNRC]…the demilitarized zone, international supervision and control" and so on. In short, the proposed changes "undermine the recognized, fundamental principle that in the South there are two administrations, two armies, two zones [of control], and three political forces." Tho and Thuy's assessment, therefore, was that the U.S. would rather continue the war than settle it. The best course of action, the envoys cabled, was to "wait out" Nixon by taking an intransigent posture at the bargaining table while letting the U.S. Congress and public pressure force him to capitulate. The VWP politburo agreed with its envoys' assessment but was not ready to scuttle the chance of getting the Americans out sooner rather than later. Instead, Tho and Thuy were to "direct the Americans to refrain from changing the Agreement" but should remain flexible enough to retract a few minor negotiating points if it helped "compel the United States to respect the Agreement…and reach a settlement as soon as possible."[69]

When negotiations recommenced on 4 December, Kissinger's task remained daunting. As Nixon's chief of staff Haldeman said at the time, "He will have to convince the North Vietnamese that if we don't get an agreement we're going to stay in, and he has to convince the South Vietnamese that if we don't get an agreement we're going to get out."[70] The U.S. and North Vietnamese delegations did little more than nibble at the edges. The Americans continued to press for changes that might nudge Thieu toward acceptance, while Tho attempted to either exact his own unacceptable concessions in exchange for those changes or simply insist on a full return to the October draft agreement. The talks soon gridlocked, with neither side willing to budge enough to coax the other to final settlement. Finally, Kissinger asked for a one-day recess to confer with Washington. "There is almost no doubt that Hanoi is prepared now to break off the negotiations and go another military round," he cabled Nixon. "Their own needs for a settlement are now outweighed by the attractive vision they see of our having to choose between a complete split with Saigon or an unmanageable domestic situation." The way Kissinger saw it, the U.S. had only two options: return to the October draft agreement or break off the talks. While he still considered the October proposal "a good one…intervening events make it impossible to accept now." It would represent a massive propaganda victory for Hanoi while causing severe political problems at home. After all, how would the administration explain dragging its feet for nearly two months, only to return to the original draft agreement? Moreover, if the U.S. kowtowed to Hanoi, "It would deprive us of any ability to police the agreement, because if the Communists know we are willing to swallow this backdown, they will also know that we will not have the capacity to react to violations." Finally, an "honorable peace" required that there be no appearance that Thieu had been abandoned. "After all our dealings with Saigon and his insistence on some changes these past weeks, this [returning to the October draft] would be tantamount to overthrowing Thieu. He could not survive such a demonstration of his and our impotence."[71]

Therefore, Kissinger recommended breaking off the talks and resuming a concerted bombing effort to bring Hanoi around. The politburo has "apparently decided to mount a frontal challenge to us such as we faced last May," he continued. "They are gambling on our unwillingness to do what is necessary; they are playing for a clearcut victory through our split with Saigon or our domestic collapse."[72] Harris polls in early December showed Nixon with a 64 percent approval rating, a record high for 1972. Moreover, 63 percent agreed that the president was doing all he could to end the war, 59 percent approved of his handling of Vietnam, and more than half believed that Nixon should not sign any agreement that failed to address South Vietnam's concerns. Given such numbers, Kissinger suggested the president go on national television to ask the American people to support another round of bombing and continued negotiation. "We have a strong case," he wrote.[73]

But Nixon was not yet ready to walk away. He ordered Kissinger not to return to the October draft but to continue pressing Tho for changes. Nevertheless, Nixon admonished his envoy to "make the record as clear as possible…that the responsibility for the breakdown rests with the North Vietnamese…that they have reneged; first as to the meaning of the agreement on the political side by reasons of the translation problem and second because they have insisted on maintaining the right of North Vietnamese forces to remain permanently in South Vietnam." Even so, the president declared that he was ready to order "a very substantial increase in military action against the North, including the use of B–52s over the Hanoi-Haiphong complex," and he was prepared to do so within 24 hours. Still, Nixon preferred achieving an acceptable agreement now rather than hoping a resumption of the bombing would net a better one later.[74]

By the end of the 7 December session, it was becoming clear

that any settlement, regardless of provisions and language, would likely not last long. In his summary of the day's action, Kissinger reported, "It is now obvious…that they [Hanoi] have not in any way abandoned their objectives or ambitions with respect to South Vietnam," he warned. "We will probably have little chance of maintaining the agreement without evident hair-trigger U.S. readiness, which may in fact be challenged at any time, to enforce its provisions." Kissinger again laid out the options as he saw them. The U.S. could either scuttle the talks and resume bombing, or go ahead with a deeply flawed agreement so long as the administration was "prepared to react promptly and decisively at the first instance of North Vietnamese violation." Kissinger later confided to Haldeman that he preferred option two, as it would give the U.S the high ground when responding to North Vietnam's all-but-certain violation of the settlement. Though fully aware of his ongoing "Thieu problem," Nixon again agreed that the risks of resuming the bombing were much greater than those of attaining an unsatisfactory agreement. He ordered his envoy to press forward, with the only condition being that "the agreement we get must be some improvement over the October agreement."[75]

What neither man knew at the time was that the politburo believed its hand was growing stronger by the day. In response to Tho's report on the 8 December session, Hanoi assessed that, "Even though the U.S. is being forced to withdraw from the war in Vietnam, they still want to achieve the best possible settlement for the U.S. and their puppets." Therefore, the politburo admonished, "We will not agree to any settlement that includes anything that might be interpreted as stating that South Vietnam is a separate country. This includes such wording as, '... the four countries of Indochina,' '… within the territories of North and South Vietnam,' etc. We must continue to demand the withdrawal of U.S. civilian personnel because this is an important aspect of ending U.S. involvement in South Vietnam."[76]

Over the next two days, the North Vietnamese delegation put Kissinger through the wringer. Haig later described the atmosphere as "brutal" and marked by "the incalculably frustrating tactics…used by the other side." Kissinger reported that the communist delegation's approach comprised "equal parts of insolence, guile, and stalling." Tho displayed no sense of urgency, Kissinger continued, and given the immense of amount of work still remaining for the 12 December session, seemed to be stalling to ensure a "sloppy conclusion, which is precisely one of their favorite tactics." The American speculated that, "Hanoi may well have concluded that we have been outmaneuvered and dare not continue the war because of domestic and international expectations. They may believe that Saigon and we have hopelessly split and that the imminence of Christmas makes it impossible for us to renew bombing the North." If that were the case, Kissinger believed "a total collapse by us now would make an agreement unenforceable. The President must also understand that an agreement at this point and under conditions that led to the collapse of South Vietnam would have grave consequences for his historic position later." Nevertheless, the envoy resolved not to break off talks, instead suggesting that they continue via diplomatic channels and that he return to Washington.[77]

Replying for Nixon, Haig concurred and advised Kissinger to suggest a recess at the conclusion of the next day's session. If there was no movement, Kissinger was to return home. If the envoy believed a "workable settlement" could be finalized, he should continue the work in Paris. Meanwhile, the president was "fully prepared to react strongly and to weather through a continuing intransigent position by Hanoi." To that end, he ordered preparations to "reseed the mines tomorrow and be prepared to move immediately with around-the-clock bombing of the Hanoi area." Haig left it to Kissinger's discretion whether to inform Tho of the latter.[78]

Despite their negotiating tactics, Tho and Thuy recommended that Hanoi take the agreement ahead of the 13 December session. If not, "it is possible that the talks may be suspended for a period of time and the war will continue." Although time was running out for the Americans, the envoys warned it is likely "they will make massive concentrated attacks for a time and then request resumption of the talks." Such attacks would surely result in "additional losses in North Vietnam, losses that will have at least some effect on the situation in South Vietnam." Though time was on Hanoi's side, they continued, "We need to recognize our opportunity. Right now the U.S. needs a settlement, but if we leave things too long…our pressure on them will have little effect, because everything has limits." Nevertheless, the politburo rejected the advice and directed the delegates to maintain a hardline stance.[79]

The North Vietnamese delegation responded by reintroducing a host of Hanoi's previously dropped demands that seemed deliberately designed to scuttle the talks. These included a resurrected insistence that political prisoners be released on the same timetable as the POWs, retaining references to the PRG—while deleting nearly a dozen mentions of the GVN, preserving language suggesting the NCNRC had "supervisory" authority over Thieu's government, and so on. Tho ended the 13 December session by asking for a suspension in the talks so he could return to Hanoi. He suggested the two sides continue to exchange diplomatic notes but no date was set for resuming negotiations.[80]

Kissinger, his exhaustion and frustration evident, summarized for Washington his experiences regarding the 13 December session. It appeared, he wrote, that "Hanoi has decided to play for time. Their consistent pattern is to give us just enough each day to keep us going but nothing decisive which could conclude an agreement. They wish to insure that we have no solid pretext for taking tough actions. They keep matters low key to prevent a resumption of bombing. They could have settled in three hours any time these past few days if they wanted to, but they have deliberately avoided this." Instead, for every one of their supposed "semi-concessions," he continued, "they introduce a counter-demand" or agree only to move objectionable sentences and provisions from one part of the text to another instead of simply deleting as requested. Thus, he concluded, "We now find ourselves in an increasingly uncomfortable position. We have no leverage on Hanoi or Saigon, and we are becoming prisoners of both sides' internecine conflicts. Our task clearly is to get some leverage on both of them." Gaining such leverage, however, had been made exceedingly difficult by Thieu's open split with the U.S. over the settlement. "If Thieu had adopted a common position with us we would have an excellent ground on which to stand now with North Vietnam's insistence on maintaining troops in the South and total refusal to recognize any aspect of sovereignty for South Vietnam." Continuing to drag out negotiations, he went on, only makes it that much more likely that Congress would step in and cut funding for the war. And with that, "everything we have striven four years to avoid will be imposed on us." The only realistic option at this point, the envoy concluded, was to return to the military option. Kissinger departed for Washington that evening.[81]

On 14 December, NSC staffer John Negroponte produced an analysis of North Vietnam's behavior since the reopening of talks on 20 November. "Hanoi has no intention to meet any of the basic requirements that we made clear to them at the end of October; and through a series of irritating dilatory tactics has pursued a course which can be interpreted as desire to achieve either no agreement at all or an agreement substantially worse than that achieved in late October. Hanoi's tactics have been clumsy, blatant, and fundamentally contemptuous of the United States." Meeting with Haig and Nixon,

Kissinger summarized the situation. "We are now in this position: as of today, we are caught between Hanoi and Saigon, both of them facing us down in a position of total impotence, in which Hanoi is just stringing us along, and Saigon is just ignoring us. I do not see why Hanoi would want to settle three weeks from now when they didn't settle this week. I do not see what additional factors are going to operate."[82]

To break the gridlock, momentum built for a new, all-out air campaign to unleash unprecedented fury upon the North Vietnamese heartland. Haig, at this point Nixon's most trusted adviser on Vietnam, favored a settlement. But when none was forthcoming, his became the most convincing voice for escalating the bombing. He later wrote that, "If we were going to strike the enemy, then we should strike hard at its heart and keep striking until the enemy's will was broken. Any operation that sacrificed lives but did not do the job was morally indefensible."[83] Kissinger agreed. But if no agreement was at hand after two weeks of "bombing the bejeezus out of them," he said, the U.S. should simply go its own way. Nixon directed JCS planners to begin preparing a comprehensive package.[84]

The president decided to give Thieu one last chance, tapping Haig to lead the charge. The general arrived in Saigon on 19 December and presented Thieu with a letter from the American president. After a preamble asking for trust and unity in seeking a peace settlement, Nixon laid out in "brutally frank" terms the choices confronting the RVN president. Haig had not come to negotiate, Nixon wrote. Instead, Thieu's choice was either to accept "this absolutely final offer on my part…or to go our separate ways." If Thieu opted for the latter, it "would be an invitation to disaster" that would "result in a fundamental change in the character of our relationship," a thinly veiled threat to cut off all U.S. support. As soon as Hanoi met his minimum requirements for peace, Nixon wrote, he would sign the agreement with or without

Saigon. He concluded by admonishing the South Vietnamese president to "decide now whether you desire to continue to work together or whether you want me to seek a settlement with the enemy which serves U.S. interests alone."[85]

After summarizing the latest draft from the 12 December session, Haig restated the U.S. ultimatum: "Under no circumstances will President Nixon accept a veto from Saigon in regard to a peace agreement." Although stunned by the enormity of it all, Thieu peppered Haig with questions about various provisions, including whether NVA troops would be permitted to remain in his country. Prohibitions on further infiltration and agreements on demobilization, assured Haig, would mitigate the danger as long as *Hanoi abided by the agreement.* After some moments, Thieu said: "It is obvious that there will be no peace as a result of this agreement," he scoffed. "After the cease-fire, the enemy will spread out his troops, join the Viet Cong, and use kidnapping and murder with knives and bayonets." Thieu predicted the settlement would pave the way for a communist takeover within six months. He eyed the American. "What I am being asked to sign is not a treaty for peace but a treaty for continued U.S. support." The general nodded.[86]

The next day, Thieu kept Haig and Bunker waiting five hours while he conferred with advisers. The South Vietnamese president marveled at how the fate of 600 American POWs had "paralyzed" a nation as powerful as the United States. Nixon, he now realized, would never send U.S. forces back to Vietnam, no matter how egregiously Hanoi violated any agreement. The American president would not risk creating a new batch of POWs. Thieu had spent the preceding weeks fretting over his predicament. If he bowed to Nixon's demands, popular resentment among both the public and his generals would boil over, threatening his government's survival and making a communist victory more likely. But if he refused to sign, the U.S. president would abandon his country, all but guaranteeing Hanoi's eventual triumph, as well.[87]

He handed Haig his own letter. Thieu began by thanking Nixon for all he had done to advance the cause of freedom in South Vietnam, recapitulating the progress the two had made in their four years together. But the communists' demands were "unreasonable" and threatened to upend all they had striven for. It was therefore unfair to force South Vietnam to either accept such an agreement or have aid from its "principal ally" cut off, especially in the face of a "ruthless enemy…aided by the entire Communist camp…" who had not abandoned "his aggressive and expansionist designs."[88] Nevertheless, Thieu understood Nixon's need to finalize an agreement. As a gesture of "maximum goodwill and as the very last initiative," he therefore pledged to accept the 12 December draft— provided any language portraying the PRG as a legitimate governing authority be stricken from the text and that all NVA troops be withdrawn from South Vietnam. These demands were inviolable, Thieu wrote, "because there can be no self-determination unless all the Communist aggressors leave South Viet Nam in fact as well [as] in principle." The Vietnamese parties could then negotiate all other issues among themselves. This, he declared, was as far as the RVN could go. Anything more would be "tantamount to surrender." He pleaded that the U.S. push for these changes with "vigor and conviction." If the communists still rejected settlement, then international and domestic public opinion "will realize better who is the obstacle to peace."[89] Haig, aware that both demands were non- negotiable for Hanoi, interpreted Thieu's response as a rejection. In his report to Washington, he wrote that Nixon now had two choices. Sign an unacceptable agreement or simply admit defeat and withdraw U.S. forces in exchange for the POWs' release.[90]

After hearing of Thieu's rejection, a frustrated Nixon

contemplated simply signing a separate agreement with Hanoi and getting out. It would likely mean the end of South Vietnam, but the president was at the end of his rope. It was Kissinger, though deeply resentful of Thieu, who dissuaded him. After all, as Kissinger would later write, "We could not in all conscience end a war on behalf of the independence of South Vietnam by imposing an unacceptable peace on our ally."[91] Rather than going it alone, he counseled, hit the North hard and then ask for a new round of talks in early January. The U.S. could then press for the best terms it could get to ensure RVN survival, but Thieu would be cut out of the process entirely. It was the least-worst compromise wedged between two equally unappealing options. Nixon agreed. He had done his own soul- searching in the days preceding Haig's visit to Saigon. After sacrificing so much American blood and treasure over the past decade, an "honorable peace" required that South Vietnam not be abandoned. That meant securing an agreement for "a peace that would last," not one that may "bring peace now, but plant the seeds for war later." The escalated bombing, he hoped, would convince Hanoi to return in good faith to the bargaining table. Nixon later recalled that the determination to again risk American lives over North Vietnam was "the most difficult decision I made during the entire war."[92]

The politburo had gambled that time was on its side, that the longer the peace talks dragged on, the more likely it was that Nixon would cave to domestic pressure. But as Kissinger would later recount, Hanoi had committed a "cardinal error. Nixon was never more dangerous than when he seemed to have run out of options." Determined not to have his second term roiled by Vietnam, the president and his inner circle saw two options: "take a massive, shocking step to impose our will on events and end the war quickly, or let matters drift into another round of inconclusive negotiations, prolonged warfare, bitter national divisions, and mounting casualties."[93]

On 15 December, Nixon directed Col. George Guay, the military attaché in Paris, to present Hanoi with an ultimatum: accept the 23 November draft as the basis for resumed peace talks on 26 December, or U.S. airpower would again train its sights on North Vietnam's military-industrial heartland. If the president received no answer with 72 hours, Hanoi would bear the responsibility for the consequences.[94] In the preceding weeks, Nixon had ordered preparations to re-mine North Vietnam's harbors and ports. He also directed JCS Chairman Adm. Thomas Moorer to work up a comprehensive plan for hitting strategic and military targets above the 20th parallel, with a concentration on the Hanoi-Haiphong complex. The commander-in-chief did not mince words. "I don't want any more of this crap about the fact that we couldn't hit this target or that one," he thundered. "This is your chance to use military power to win this war. And if you don't, I'll hold you responsible."[95]

Moorer tapped Strategic Air Command (SAC) chief Gen. John Meyer and his Offutt Air Force Base staff in Omaha, Nebraska to plan the new campaign. Offutt had actually been preparing for a more expanded role in Route Pack VI since late summer. By the time Moorer came calling, planners had already developed an initial target set of some 60 locations in Route Pack VI. These included airfields, POL facilities, power plants, railyards, military warehouses and transshipment points. Aside from their strategic value, planners chose targets that were sufficiently large and discrete to minimize collateral damage.[96] On 15 December, the JCS issued a warning order to Meyer and PACOM commander Adm. Noel Gayler: be prepared to launch a "maximum-effort strike" at any time.[97]

The weapon of choice would be the B-52 Stratofortress. The bomber had been used to great effect south of the 19th parallel but had never been employed in large numbers over North Vietnam's heartland. And with good reason. Planners were worried the slow and sluggish BUFFs would be especially vulnerable to the massed air defenses of Route Pack VI. But with the impasse in Paris, that was about to change. The big bomber brought several advantages. December cloud cover promised to be particularly bad this season. The all-weather B-52's advanced avionics and navigation systems could cut through that obstacle with relative ease. More importantly, the strategic bomber was an instrument of unparalleled shock value. Cruising at 30,000 feet, a single B-52 carried a 30-ton payload of 500 and 750-pound bombs. The psychological impact of wave after wave of three-plane cells raining devastation upon mile-wide swaths could scarcely be overstated. Adding to that impact was the campaign's relentlessness. Designed as an around-the-clock operation, the new campaign called for mass B-52 night attacks, along with highly selective F-111 and A-6 raids on airfields, communication hubs, and air defense sites. Meanwhile, hundreds of 7th Air Force and Task Force 77 tactical fighter-bombers were to pound targets during the day. Planners dubbed the new effort Operation Linebacker II.[98]

Over the course of 1972, Operation Bullet Shot and its successors had steadily grown the B-52 and KC-135 Stratotanker fleet in Southeast Asia. By December, the force had swelled to some 206 B-52s, or about half of SAC's existing bomber inventory. The 307th and 310th Strategic Wings (SW) and their 54 B-52Ds were stationed at Thailand's U-Tapao Air Base under Brig. Gen. Glenn Sullivan. Another 53 Delta models of the 43rd SW and 99 B-52Gs of the 72nd SW under Brig. Gen. Andrew Johnson were positioned at Andersen Air Force Base, Guam. All B-52 operations in Southeast Asia were overseen by Eighth Air Force commander Lt. Gen. Gerald Johnson from his headquarters on Guam.[99]

That tiny isle had been especially overwhelmed by the influx of men, materiel, and machines. Colloquially known as "The Rock," the 212 square-mile island had been acquired from Spain as a spoil of the 1898 Spanish-American War. Located about 1,500 miles east of the Philippines, Guam had by the mid-1960s become a major staging area for B-52 Arc Light missions in Indochina. Over the previous nine months, its B-52 force had simply exploded, rising from about 40 planes to more than three times that number by the eve of Linebacker II. One of Andersen's runways had to be closed to make room for the added aircraft, and an additional maintenance wing had been stood up in July. More than 12,000 airmen thronged the facility, overwhelming the base's infrastructure. The enlisted were booted from their concrete barracks to make room for the influx of B-52 crews. The men soon found themselves crowded into makeshift "tin cities," clusters of unairconditioned 1950s-era steel huts. Originally designed to house 96 men in four bays, twice that number were now shoehorned into the egg-shaped structures, with just four toilets and urinals and another four showers to share among them. And they were the lucky ones. Those who arrived later in the year were forced to live in canvas tents, often crawling with rats and the island's ubiquitous brown tree snakes which had slithered in to feed on them.[100]

Air Force avionics tech Sgt. Steve Williams was one of the "lucky" ones. Deploying in early April, Williams spent his first night sleeping on the gym floor before landing a spot in Tin City. The Lexington, MA native then spent a sweltering summer trying to keep the big bombers' fragile communication gear operating. "It was a miserable place, hot and humid all the time," he says ruefully. "There was a lot of tension in the barracks because of the overcrowding. If a guy was sitting on the toilet too long, somebody would start banging on the wall. People got pissed off at each other." For the most part, work was the only distraction. And there was plenty of it. Despite its place as the United States' frontline strategic bomber, the B-52's communications gear was downright antiquated. Still reliant upon tubes and other fragile internals, the ARC-34 radios were constantly breaking under the stress of increased mission tempo and the heavy vibrations thrown off by the BUFF's big engines. Twelve-hour shifts were the norm and days off about as rare as snowfall on the tropical island. Most were spent ferrying up and down the flightline on a step van nicknamed the "bread truck," making last minute repairs while impatient BUFF crews stood by to launch.[101]

Sometimes the pressure boiled over. One particularly harried day, Williams was struggling with an ARC unit when something slipped, ruining the whole repair. "I just started cussing…and I was an NCO, so I knew how to swear," he says chuckling. Then a voice over his shoulder. 'Is all that helping you get the job done, sergeant?' Williams turned to find none other than Gen. John D. Ryan, Air Force Chief of Staff, staring him in the face. With Linebacker heating up, the service chief had flown in to take the grand tour. The general's aides shot Williams a collective look: *nice going, dumbass.* The NCO snapped to attention. "Yes, sir," he said, not missing a beat. "Helps relieve the tension so I can get it right the next time." Well, said Ryan, "As long as you keep this stuff operational, you can swear at it all you want." Williams got his fair share of ribbing over the "Ryan incident," but took it all in stride. Despite the miserable living and working conditions, he remains proud of his service. "It's the comradery of the military, of the people that depend on you to do your job. That's where the pride comes from." Williams would escape Guam that October, redeploying to Kadena to work on KC- 135s during Linebacker II. He would go on the earn his Air Force commission, retiring a captain in 1991.[102]

All told, the array of U.S. heavy bombers, strike fighters, and various support aircraft would top 1,300 by the start of the campaign. They would be flying into the teeth of a revitalized North Vietnamese air defense system. Since the 23 October bombing halt above the 20th parallel, Hanoi had busily repaired and restocked much of what had been hit during Linebacker. By the eve of the new campaign, North Vietnam could boast some 2,300 SA-2 missiles, with 200 launch sites ringing Hanoi and Haiphong alone. With an effective ceiling of some 60,000 feet, SAMs would prove the principal threat to the big bombers. Meanwhile, hundreds of radar- controlled 85 mm and 100 mm antiaircraft guns encircled targets throughout the military-industrial heartland. Though they posed little threat to the high-flying B-52s, such weapons would produce a murderous wall of antiaircraft fire for tactical strike fighters. Finally, North Vietnam's ground-vectored MiG fleet still lurked. Though decimated over the previous six months, Hanoi still had about 80 MiG-19s and 21s at its disposal.[103]

As the deadline approached, commanders received the official order: "You are directed to commence at approximately 1200Z, 18 December 1972 a three-day maximum effort, repeat maximum effort of B-52/TAC Airstrikes in the Hanoi/Haiphong areas. Object is maximum destruction of selected military targets. Be prepared to extend operations past three days, if directed." The order further admonished commanders to "exercise precaution to minimize risk of civilian casualties."[104] When Nixon's 72-hour ultimatum came and went, he ordered the first of 129 B-52s to launch late on the afternoon of 18 December. Although Hanoi would later claim it had not received Nixon's message until after the bombers were airborne, the point was moot. Operation Linebacker II had begun.[105]

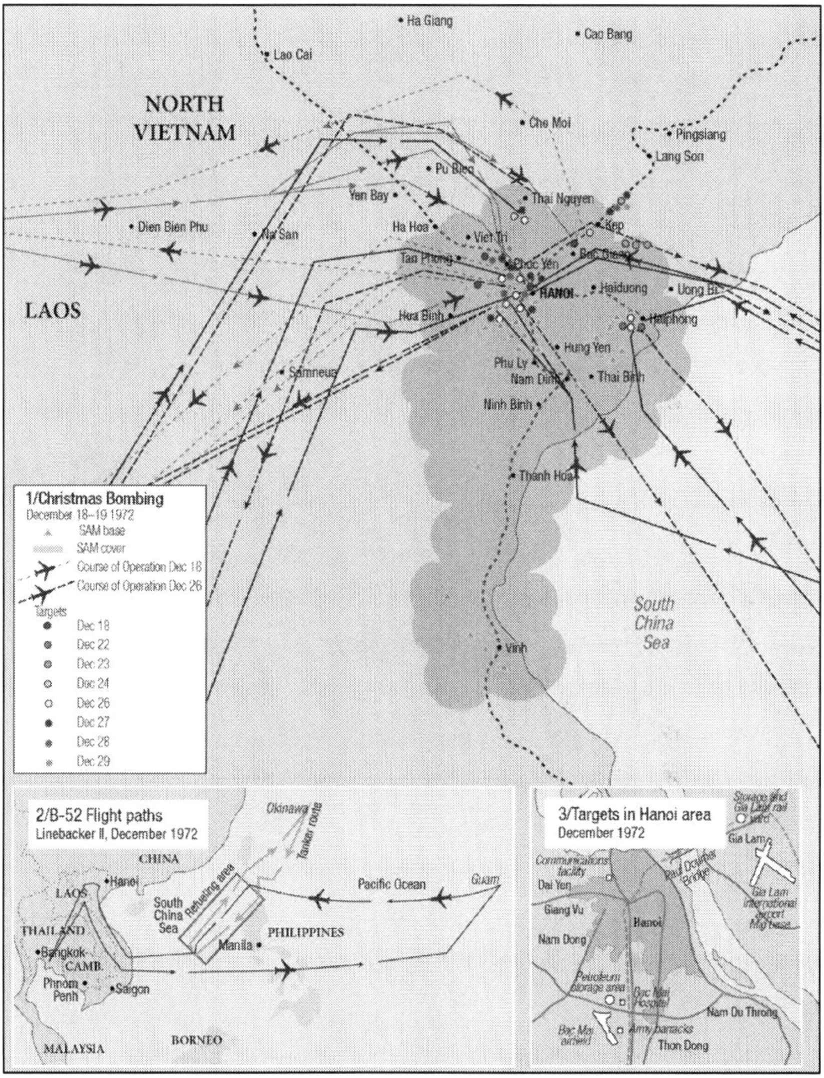

Figure 4: Operation Linebacker II

9

A MAXIMUM EFFORT

December 18 – 29 January

"Gentlemen, your target for tonight is Hanoi," said Col. J.R. McCarthy, commander of the 43rd Strategic Wing at Guam's Andersen Air Force Base. An illuminated map of North Vietnam glowed on the slide screen behind him, a large triangle superimposed over the capital city. It was morning, 18 December, and the room was packed with more than 100 B-52 airmen who in a few hours would fly the first waves of massive strategic bombers into the teeth of the world's most heavily concentrated air defense system. Planners estimated losses at 10 percent. A silent pall of anxiety and intense concentration settled over the crowd. McCarthy, who had seen more than a thousand such briefings over his career, marveled at the sight. "I can truthfully say that that group of combat crews was the most attentive I have ever seen," he later wrote.[1]

Capt. Robert "Bob" Certain, 25, a navigator with the 340th Bomb Squadron, was in the audience that morning. Ironically, he and his crew had been scheduled to rotate back to their Blytheville, AR base two weeks earlier. But a stateside snowstorm had waylaid their replacement crew, and now Certain's B-52G would be among the first to go in. The Savannah, GA native had already flown more than 50 missions over South Vietnam and the North's southern panhandle. But other than "the copilot spilling hot coffee on himself during the bomb run," flying those "friendly skies" left very little to worry about for the high-flying BUFFs.[2] Hanoi was a different story. Swarms of SA-2s, and even the North's dwindling MiG fleet, would pose clear and present dangers above the 20th parallel.

The inaugural attack would come in three waves. The first would see 18 B-52Gs and nine B-52Ds from Andersen make the eight-hour flight to a rendezvous point over northeastern Laos. There they would link up with another 21 Delta models out of U-Tapao before turning east toward Hanoi. Calling to mind the World War II bomber formations of old, the BUFFs would then form up for an "elephant walk," a continuous line of three-plane cells stretching some 70 miles, all at the same altitude, and spaced about 10 minutes apart. In keeping with SAC protocol, each cell was identified by a color-coded call sign. Their targets were the airfields at Hoa Lac, Kep, and Phuc Yen, along with the Kinh No and Yen Vien railyards just north of the capital.[3]

The Deltas had been flying Arc Light strikes in Southeast Asia since 1965 and boasted a massive 30-ton conventional payload. Though older, the D-models had been upgraded under the 1966 Big Belly program with improved electronic and ECM systems. The Golfs, on the other hand, had only recently been rushed into theater following Hanoi's Nguyen Hue Offensive that spring. The newer G- models were faster and had a longer range—valuable characteristics for a strategic nuclear bomber— but they packed just a third of the Delta's conventional payload. Most crucially, the Golfs often lacked the latest electronic and ECM technology, a real liability in a crowded battlespace dominated by radar-guided air defenses. Both models equipped a manned, radar-guided tail gun: the Delta a quad- 50-caliber machine gun, the Golf a 20 mm cannon. Each was crewed by a pilot/aircraft commander, copilot, radar navigator/bombardier, navigator, electronic warfare officer (EWO), and tail gunner.[4]

As with Linebacker I operations, an array of support aircraft was crucial to mission success. F-4 chaff Phantoms, Wild Weasel and Iron Hand air defense suppressors, and fighter escorts led the way, while EB-66 radar jammers, early warning and communication E-2B Hawkeyes, and refueling tankers stood off to provide support. KC-135s out of Kadena Air Base, Okinawa were an especially crucial asset. For the B-52s out of Andersen, a preset refueling station was set up west of Luzon, Philippines. There, the 135s would top off the BUFFs for their 17-hour, 5,200-mile round trip to Hanoi. The big Stratotankers would also assume a large share of the refueling responsibilities for other Linebacker II aircraft.[5]

Several flight aborts before takeoff had pushed Certain's BUFF a few spots up to the cell's lead position. And that was a good thing. Every spot forward upped your chances of survival since enemy ground crews had less time to draw a bead on your aircraft. Still, Certain and his crewmates remained the 10th G-model in line...but it was better than being the 12th. Now designated *Charcoal 01*, the bomber was piloted by Lt. Col. Donald Rissi, 41, from Collinsville, IL. Next to Rissi was copilot Lt. Robert Thomas of Madison, GA, who in less than 12 hours was due to celebrate his 24th birthday. San Francisco native Maj. Richard Johnson, 36, would handle radar navigation/bombardier duties. Capt. Richard T. Simpson, 31, hailed from Anderson, SC and served as the ship's EWO. With Certain as navigator, Sgt. Walter "Fergie" Ferguson, 43, rounded out the crew. The Hope, AR native was tasked with manning the G-model's radar-guided 20 mm tail gun and for issuing visual missile warnings from his bubble canopy.[6]

Just after 1900, the long train of bombers turned east out of Laos and crossed into North Vietnam. In Nghe An Province, operators of the 45th Radar Company, 291st Regiment watched as the big bombers rolled in. At 1910, station commander Nghiem Dinh Tich alerted Air

Defense Headquarters: "B-52s are now flying toward Hanoi." Within five minutes, the politburo's Central Military Party Committee (CMPC) had issued a nationwide alert. Fearing a renewal of American bombing when the peace talks stalled in late November, the politburo had ordered the Air Defense-Air Force Service to resume preparations for heavy strikes above the 20th parallel. This included in-depth study of B-52 tactics and ECM capabilities. The results were disseminated to antiaircraft batteries throughout the region, as was additional SAM equipment, SA-2 stocks, AAA weapons and ammunition. Emergency rescue equipment and supplies were pre-staged, and repair and construction of bomb shelters accelerated. Supply stocks, especially those likely to be targeted, were dispersed wherever possible. Finally, the politburo ordered children, elders, and others not directly contributing to the economy or war effort evacuated from Hanoi and Haiphong. By 18 December, some 200,000 nonessential personnel had been relocated to the countryside.[7]

Rissi clicked the interphone. No need to make threat missile calls, he said. Just focus on putting bombs on target. It would do no good anyway. This was a "press on" mission, and crews were ordered to take no evasive action to avoid enemy missiles or fighters. This meant maintaining straight and level flight in a high-threat environment for at least four minutes prior to bomb release. Primarily, CINCSAC had ordered these highly restrictive rules of engagement to minimize civilian casualties. Maintaining steady level flight also helped the bombardier's accuracy, maximizing target destruction while helping limit collateral damage. Finally, planners believed that maintaining cell integrity also enhanced mutual ECM support. The tactic would soon be put to the test. Frantic calls from leading cells crackled over the net as they began picking up masses of SAM radar signatures. It was "wall to wall" SAMs up there, someone radioed, and *Charcoal 01* would be in lethal range within minutes. CINCSAC's orders, recalled Certain, began to look "increasingly suicidal" by the minute. "It was insulting because it seemed that the safety of the enemy was taking precedence over our own." Certain started a mental checklist of all the tactical changes he would recommend once back at Andersen.[8]

About 101 nautical miles from target, the bomber train split into two elements. *Charcoal 01* and another 26 BUFFs continued straight for the railyards at Kinh No, Yen Vien, and the storage facilities north of the capital. The remaining 21 aircraft turned to hit the airfields at Hao Lac, Kep, and Phuc Yen. The targets had already been struck earlier, when just after sunset a group of 16 stealthy F- 111s from the 474th TFW out of Takhli swept in to dump 50 tons of high explosives on the MiG bases. Repeated strikes on the same target was to become a common theme over the course of the operation, as planners sought to maximize destruction.[9]

A few minutes more and Certain's group split again, with *Ebony, Ivory,* and *Charcoal* banking southeast to concentrate on Yen Vien. *Charcoal* would be last in line over the target. "Radar, where's the chaff?" Certain called. During the preflight briefing, crews had been warned that the aluminum chaff corridors laid by preceding F-4s might make it difficult to locate the target on radar. Certain was picking up the target just fine. But no chaff cloud. Johnson tuned through several frequencies before finding it. Incredibly, the Phantoms had laid the corridor from the turn to the target. On the scope, it looked like a flashing neon sign for SAM crews to follow. *Why didn't they just send the route maps to Hanoi?* wondered Certain ruefully. *It looks like our own Air Force has set us up!* Another worry was the post-strike turn. Following bomb release, *Charcoal* was to make a long, sweeping turn to the west to exit the target area. But that turn would be straight into the face of a crisp headwind, cutting the bomber's speed by nearly 200 mph—never a good thing in a high-threat environment. The turn would also blank out their SAM radar jammers for nearly two minutes, making the big bombers "visible" to ground crews. It was just "another insane feature of our worsening tactical situation," Certain recalled. His mental checklist was growing longer by the minute.[10]

As Rissi banked toward the target, EWO Simpson reported a radar lock by 100 mm AAA—nothing to worry about at 30,000 feet. No SA-2 activity, either, he noted. But what the crew did not know was that their B-52G had been rushed into theater without the latest ECM technology. While the D-models had been upgraded to the newer and more effective AN/ALT-22 Modulated Transmitters, their BUFF was still using the older ALT-6B Unmodulated Transmitters—equipment SAM crews had learned to defeat. Using the so-called "three point technique," two radarmen working in concert could identify the ALT-6B's jamming strobe on their Spoon Rest acquisition radar. Switching on the Fan Song guidance system sometimes allowed crews to use the first two signals to triangulate the enemy plane's radar return and acquire a tracking solution. Now, as *Charcoal 01* approached 30 seconds to bomb release, its crew had no idea that it was *they* who were in the crosshairs.[11]

Below, technicians with the 59th Missile Battalion locked on and fired a pair of SA-2s at *Charcoal 01*, part of a 17-missile volley aimed at the cell. Less than a minute later, both erupted near the G-model, shredding the cockpit and fuselage with flaming hot shrapnel. Other fragments were sucked into the four port engines, ripping their internals and knocking them out. A violent shudder rumbled through the aircraft as Certain felt the BUFF yaw to the left. At the same time, the radar screens blacked out. The navigator's first thought was that someone had accidentally knocked the generators offline. Then an urgent call on the interphone from Thompson, the copilot: "They've got the pilot! They've got the pilot!" Then Simpson's voice broke through.

"Is anyone there?" called the EWO. "Gunner, gunner!" Simpson's station had gone black, electrical shorts snapping and popping in the darkness. Ferguson, the tail gunner, could not answer. He had been killed as a shower of flaming metal shards sliced through his firing station at 8,000 feet per second. Certain tried to wrap his mind around what was happening. *This can't be. We haven't been hit, or have we?* All at once, the 25 year old accepted the situation, his training taking over. Through a porthole window in the bulkhead door he spotted a fire blazing in the forward wheel well, a clutch of 750-pounders in the nearby bomb bay. "Drop those bombs!" he called to the bombardier. Johnson's own training kicked in. Even under extreme duress, he remembered to safety the bombs before release.[12]

Certain's mind next snapped to the main mid-body fuel tank just above the fire, its 10,000 pounds of JP-4 jet fuel inches from the flames. It was time to get out. The interphone crackled. "The pilot's alive," called Rissi, his voice weak and fading. The shrapnel had gotten him too, but he was hanging on. "EW's leaving!" called Simpson. It was every man for himself, now. Roughly 10 seconds had passed since the SAMs had hit. The ballistic activators under the EWO's seat fired, and Simpson rocketed up and out of the stricken craft. Certain needed no further encouragement. He stowed his work table, pulled his helmet visor down, and cinched his oxygen mask tight. He saw the red ejection light flash. That was Rissi blasting free. The navigator grasped the ejection handle between his knees and pulled. Nothing happened. A flood of wild panic swept over him, mind racing. *Was this thing jammed on the rails? I'll be beaten to death by the wind*! But it was all a shock-fueled illusion. Time had slowed to an imperceptible crawl. In three-tenths of a second, both his seat activators and explosive hatch bolts had blown just as they should, blasting open the hatch below and firing Certain and his seat free of the plane. It was just after 2000 hours.[13]

Freezing wind was his first sensation. The next was that he was tumbling wildly. *Got to stabilize*, his mind warned. If the parachute deployed now, it would simply wrap him in a silk death shroud as he plummeted to earth. The navigator tucked his arms and legs, letting his center of gravity stabilize. The chute deployed at 15,000 feet. He watched a dazzling train of 750-pounders from *Ebony* walk a fiery path through the railyards below. Everything seemed to be on fire. "It looked like the jaws of hell," he says. Then realization struck: he was right over the target. Every few seconds another bomb would whistle by. He was either going get one on top of his head or he would drift into the middle of that inferno below and be burned to death. "And I'm thinking, 'Holy…a guy could killed. I gotta get away from this.'" He reached up and grabbed a fistful of parachute risers, pulling the cords down with all his might, hoping to tilt the chute and let the breeze carry him away to safety. It worked. As the wind pulled him west, he spied the outline of an immense flaming arrow embedded in the earth. Fearing that he was drifting toward another target area, he at last realized it was not a target at all, but the final resting place of *Charcoal 01*. It was just the second B-52 to shot down during the war and the first of Linebacker II. It would not be the last.[14]

Certain came down in a suburban area west of the capital and was captured immediately. Local militia also nabbed Simpson before he could even get free of his parachute. Johnson, the bombardier, managed to evade some 12 hours before he too was caught. The three would be taken to the nearby Hanoi Hilton, where they became the first POWs of Operation Linebacker II. All would be released during Operation Homecoming in early spring 1973. Although pilot Rissi managed to eject, he would die at some point that night. It is unclear whether copilot Thomas was able to eject, but he too was killed, just hours short of his 24th birthday. Tail gunner Ferguson never made it out of the plane. The remains of all three men would be returned to U.S.

custody on 23 August 1978.[15]

On the heels of Wild Weasel strikes, the night's second wave rolled in just before midnight. Thirty B-52s out of Andersen followed chaff corridors laid by the Phantoms to restrike the Kinh No and Yen Vien targets, hitting the railyard and repair facility east of downtown Hanoi, as well. The raiders were again hit with a wall of SAMs, this time 68 of the big missiles rocketing skyward to down the BUFFs. Another G-model from the 72nd SW was badly hit as it was making its turn just after bomb release, setting it ablaze and knocking out two engines. Tail gunner Master Sgt. Ken Conner took shrapnel through the arm but was otherwise okay. All other crewmembers were unscathed. Incredibly, the aircraft was stable enough for pilot Maj. Cliff Ashley to continue evasive maneuvers until the plane was out of lethal SAM range. Navigator 1st Lt. Forrest Stegelin, his instruments blown, manually plotted a heading for U- Tapao. With a pair of F-4s flying escort, Ashley nursed the stricken craft across the Thai border before the crew was forced to punch out. A Marine chopper from nearby Nam Phong rescued all six crewmembers within 20 minutes of touch down.[16]

The day's largest strike, featuring 30 Andersen and 21 U- Tapao BUFFs, hit just before dawn. The Kinh No railyard was again pummeled, along with further strikes on Radio Hanoi facilities. Navy Iron Hands had taken over SAM suppression for exhausted Weasel crews, but despite vigorous efforts, the enemy managed to unleash more than 150 SA-2s. Again, SAM crews struck home, this time downing a Delta-model attached to the 307th SW out of U-Tapao.

Two crewmembers were killed and another four added to the Hilton's guestlist. Another Delta, this time with the 43rd SW out of Guam, took some damage from an exploding SAM while hitting Radio Hanoi. The D-model was also able to maneuver to Thai airspace and land safely at

U-Tapao. Day one was over.[17]

The U.S. had flown 121 B-52 sorties, along with select night strikes by F-111s, against the North Vietnamese industrial heartland, pummeling targets—some for the first time in the war—with more than 1,000 tons of ordnance. SAM and antiaircraft emplacements also felt the wrath, as Wild Weasel and Iron Hand strikes worked to degrade capabilities. As day broke over the capital, the unprecedented scale of smoking wreckage told the tale, as the night strikes had hit an astounding 94 percent of targets. While the few air- to-air clashes between F-4s and MiGs yielded no kills for either side, Staff Sgt. Samuel Turner aboard the 307th's *Brown 03* became the first tail gunner of the war to shoot down a MiG-21. He would later be awarded the Silver Star for his actions.[18]

Barry Romo, a former U.S. Army second lieutenant and Vietnam veteran who had since turned against the war, was in Hanoi along with other antiwar activists on 18 December. "The raids were worse at night," he recalled. "We would go into the bomb shelter with the workers and other people staying at the hotel. The B-52s were so loud, when they dropped bombs, you could feel the earth move and grind. The shelters, being underground, only intensified the feeling of the earth grinding. It's hard to explain what it was like in the shelters. Everyone was on edge, just wanting the bombing to end so they could get out."[19]

The day had proved costly for U.S. fliers, as well. The North Vietnamese fired some 200 SA-2s over the night and early morning, downing three B-52s and damaging two others. Another Air Force F-111A was lost to unknown causes during a night attack on the Radio Hanoi facilities. Five airmen were KIA and another seven made POWs. While the 2.3 percent loss rate was slightly better than the predicted three percent, debriefing crews voiced serious doubts over tactics.

Airmen decried the use of World War II-like formations that saw a 70-mile-long train of bombers, each flying at the same speed, altitude, along the same track, and with cells often using the same ingress and egress routes. All of this made enemy ground crews' job that much easier. "If 36 aircraft turned at a certain point to a certain heading," complained one pilot, "it does not require much of an educated guess to decide where to aim at number 37." Most worrisome was the post target turn (PTT). Under current protocols, pilots were required to make a 100-degree turn back over target, often into a stiff headwind that dramatically cut airspeed. Worse, because the BUFF's internal ECM gear faced forward, the turn negated its jamming effects, allowing SAM radar crews to get a clear picture of the big bombers wings and underbelly. Planners noted crew concerns, but no changes were made for the next night's missions.[20]

Daybreak on 19 December saw Task Force 77 begin re- mining North Vietnamese ports, harbors, and river estuaries from Dong Hoi to the Chinese border. The mines laid during Pocket Money in early May had been set to deactivate after six months, and many were now offline. Additionally, unusually powerful solar storms in early August had caused such geomagnetic disturbances that about half of the 8,000 MK-36 DST magnetic mines originally laid spontaneously detonated. The USS *Ranger* launched several waves of A-7s to begin the process at Haiphong and Hon Gai harbors. The mission came off without a hitch and would continue in coming days. One *Ranger* A-7E from VA-113 would be shot down by ground fire on 24 December, with pilot Lt. Philip Clark, 26, killed in action. By the end of the month, all targeted minefields had been successfully reseeded with no further combat losses.[21]

Full tactical strike packages were also in the works. Over the next several days and nights, aircraft from the *America, Enterprise,*

Midway, and *Ranger* hit targets throughout the Haiphong area and all along the coast. Air defenses received the bulk of the attention. Airfields at Kep and Kien An were repeatedly hit, along with SAM and other antiaircraft emplacements. A large night raid on 20 December emphasized the point as Intruders off the *America* and *Ranger* struck 10 SAM sites in the Haiphong area. Over the first 24 hours alone, the Navy racked up 53 strike sorties, achieving multiple secondary explosions and engulfing scores of targets in flame. An A- 7 Iron Hand was shot down during a duel with a SAM site south of Haiphong on 19 December, and an A-6A was lost to 37 mm ground fire during a strike on the port city's shipyards. Three naval aviators were taken prisoner. Meanwhile, Task Force 77.1, the Navy's surface bombardment contingent, was getting its licks in, as well. Operating day and night and in all weather, surface warships hit coastal targets from Thanh Hoa up to Cam Pha, including highway ferries, coastal defense sites, army barracks, storage facilities, truck parks, and bridges. Navy guns sunk several enemy patrol boats, as well. Surface warships came under repeated artillery attack. On 20 December, the guided missile destroyer *Goldsborough* was operating near Thanh Hoa when fire from North Vietnamese coastal guns struck the chief petty officer's quarters, killing him and wounding four other sailors.[22]

Air Force daylight strikes also began in earnest on 19 December. Despite adding a contingent of A-7D all-weather tactical bombers, the service was still dealing with a dearth of aircraft capable of operating in the extremely poor conditions that December. Conditions also precluded using laser-guided bombs, so Air Force planners turned to Pathfinder Phantoms equipped with long range electronic navigation beacons (LORAN) to lead conventional strikes against a variety of targets, including air fields, power plants, radio and communication sites, and railyards. First used in September 1971, the system employed multiple ground-based transmitters that could communicate with the Pathfinders' onboard systems, allowing them to fix and communicate target positions to accompanying strike fighters. Of particular interest was the Yen Bai airfield about 70 miles northwest of the capital. The MiG base happened to sit along the main ingress route for B-52s and so was repeatedly struck on the 19th and 20th by 40 Air Force A-7Ds. Hanoi's radio and communication facilities were also prime targets both day and night, with 54 LORAN-guided F-4s striking in a daylight raid. On the 21st, 16 A-7Ds and 46 radar-guided F-4s hit railyard complexes east and south of the capital, crippling heavy transportation in the city. Gradually improving weather enabled the use of precision munitions, as 8th TFW strike fighters completely knocked out Hanoi's central rail station and its downtown thermal power plant using just eight, 2,000-pound laser-guided MK-84 bombs.[23]

The second night's B-52 raids mirrored the first, as three waves descended upon the industrial heartland. As before, myriad supporting aircraft paved the way, with chaff Phantoms and Wild Weasel and Iron Hand suppression strikes preceding each raid. EB- 66 Destroyers stood off to jam enemy radar and communications, while MIGCAP F-4s prowled the skies for enemy fighters. Another round of F-111 "whispering death" strikes on Hanoi's airfields further suppressed MiG activity. But the BUFFs continued to be the centerpiece, with 27 Deltas and 36 Golfs out of Andersen, and another 30 D-models from U-Tapao thundering overhead. Once again, crews were ordered to take no evasive actions regardless of enemy activity. Col. McCarthy, commander of the 43rd SW at Andersen, even declared that any "aircraft commander who disrupted cell integrity to evade SAMs would be considered for court martial." Night two's target sets were nearly identical to the first. Yen Vien and Kinh No rail complexes again found themselves in the crosshairs, as did Radio Hanoi, the Thai Nguyen thermal powerplant, and the Bac Giang transshipment point north and northeast of the capital. While MiG and AAA activity were nearly nonexistent, North Vietnamese crews fired some 180 SA-2s, damaging two BUFFs, one severely. Its crew managed to coax the big bomber into Thai airspace before making an emergency landing at Nam Phong. Tactics— including the worrisome PTT—remained the same, as did ingress and egress routes. Because of the nearly 17-hour round trip for Andersen BUFFs, returning crews were often still landing or debriefing even as the next night's airmen launched their missions. The highly complex nature of coordinating faraway support craft made even the most minor tactical adjustments very difficult. Moreover, with no aircraft lost on the second night, CINCSAC had become confident that mission planning was solid. That assumption would prove costly.[24]

As dusk fell on 20 December, F-111s again staged limited night attacks, once more targeting air fields and communication assets. Chaff Phantoms and Wild Weasels again led the way, with EB-66s and Navy EA-6B Prowlers providing electronic countermeasures and communication jamming. Some 99 B-52s from Andersen and U-Tapao in three waves again provided the punch. The first wave of 33 BUFFs followed the now well-trod path along the Red River to hit targets in the Hanoi area. Six bombers broke off to strike the capital's main railyard and repair facility, encountering very little resistance. The remaining 27 were not so lucky. SAM crews unleashed a murderous barrage of some 130 SA-2s. Rather than engage the big bombers, MiG-21s instead shadowed the BUFFs, relaying the Stratofortresses' airspeed and altitude to ground sites. This enabled crews to "blind fire" volleys without switching on their radar and drawing unwanted attention from prowling Wild Weasels. The effect was devastating. Two BUFFs were blasted from the sky, while a third limped into Thai airspace before crashing.[25]

Aware of the B-52G's ECM deficiencies, commanders frantically aborted a second-wave attack on Hanoi's main railyard by six G-models. The remaining 21 bombers pressed on to again hit the Thai Nguyen thermal power plant north of the city. The Bac Giang railyard to the northeast was also restruck. All second-wave bombers emerged unscathed. Just prior to dawn, the third and final wave of 39 BUFFs began its runs on Hanoi's Gia Lam rail complex, POL storage facilities, the Bac Giang transshipment yard, and, once again, the Kinh No railyards. SAM crews again struck pay dirt, sending the burning hulks of two more B-52Gs careening earthward, killing nine airmen and adding four more inmates to the Hanoi Hilton. A Delta model with the 43rd SW was also badly struck. Its crew managed to steer the crippled ship into Laotian airspace before punching out. Five crewmen were eventually rescued, a sixth listed as MIA. It was a bad end to an even worse day. In all, the Eighth Air Force had lost six of its strategic bombers, including four of the ill-equipped G- models. Two more BUFFs had been badly damaged. Coupled with the loss of life and those taken prisoner, 20 December would soon become known as "Black Wednesday" among SAC crews.[26]

Operation Linebacker II had so far cost nine B-52s lost, several more support aircraft shot down, more than a dozen airmen killed, and nearly as many made prisoners. A combination of predicable and dangerous U.S. tactics—including the ill-advised post target turn—deficient ECM equipment in the Golf models, and the adaptability of North Vietnamese ground crews had combined to teach missions planners a series of hard lessons. Nixon, already under increasing domestic pressure for resuming the bombing, "raised holy hell about the fact that [B-52s] kept going over the same targets at the same times" and ordered the problem fixed posthaste. With the nation's frontline strategic bomber in harm's way, CINCSAC Gen. John Meyer realized change was needed and quickly. First, Meyer ordered all operations consolidated under Eighth Air Force commander Lt. Gen. Gerald Johnson at Andersen. Next, if SAC were to protect its bombers and crews, degrading SA-2 missile supplies and launchers had to take priority. Post-strike reconnaissance photos had shown SAM sites running low on replacement missiles, so Meyer directed Johnson to focus on SAM sites and storage facilities, especially those along the inbound and outbound flight corridors. Planners would also need to revise mission tactics, including routing, timing, support coordination, enhancing mutual ECM support, and more. Of particular concern was the vulnerability of the ill-equipped Golf models. Fortunately, a preplanned shift in mission tempo from Andersen—which housed the bulk of the Gs—was already on the books. Since Nixon had decreed there would be no letup on the bombing, it would be up to U-Tapao's 307th and 310th SWs fleet of Deltas, along with Air Force and Navy tactical bombers, to carry the momentum. In the meantime, Eighth Air Force planners would have a short reprieve to figure out how to get the Golfs responsibly back into the fight.[27]

The 21 December night raids were substantially more focused. Just 30 U-Tapao B-52s were sent in a single wave against three brand new targets near Hanoi, including those thought to house SAM replacement stocks. Six Deltas were to hit Quang Te airfield, another 12 targeted the Ven Dien supply depot, and the remaining 12 were to strike the Bac Mai airfield and storage area.[28] Planners did their best to incorporate aircrew feedback on the fly. Whenever possible, routes, altitude and intervals between attacking cells would be varied to make it harder on antiaircraft crews. The PTT was also dramatically shortened and even eliminated whenever possible. In some cases, bomber crews over Hanoi were allowed to continue east through the high-threat areas toward the coast rather than turning back over target and heading west. There were other welcome changes. Since the BUFFs were coming from nearby Thailand, no aerial refueling was necessary and mission length was cut to just four hours. Lighter fuel loads meant increased bomb capacity for each plane, while eased logistical requirements allowed planners to add more support aircraft. For the first time in the campaign, nighttime Hunter-Killer flights were dispatched to further degrade the North's still formidable SAM defenses. As during Linebacker, H-K missions saw pairs of F-4s armed with cluster bombs working in tandem with Shrike-equipped F-105Gs to seek out and destroy SAM radar and launch assets. Higher numbers of chaff Phantoms were also added to provide wider and more effective dispersion.[29]

Once more, the BUFF strikes were preceded by F-111 night raids on various targets, including Hanoi's surrounding airfields, rail facilities, and transshipment complexes. Next came the B-52s. The Quang Te and Ven Dien waves were completely successful, all bombers delivering their payloads on target and escaping unscathed despite massed SAM launches. The Bac Mai group would not be as lucky. First, *Scarlet 01*'s bombing radar failed, necessitating it move to the back of the formation. Then a second *Scarlet* BUFF's ECM jammer foundered, rendering the plane completely visible to SAM crews. SA-2 gunners immediately let loose with a four-missile volley. The first two missed, but the second pair struck home, blasting *Scarlet 03* from the sky. Leading the next cell in line was *Blue 01,* which had just begun releasing its payload. Suddenly, it too was hit by SA-2 missiles. Between the two, nine airmen were killed and another three taken prisoner. The SAM hits on *Blue 01* also had the additionally tragic effect of throwing off its bomb train, striking nearby Bac Mai hospital and killing 28 staff members. The death toll could have been much higher had authorities not evacuated patients prior to the start of Linebacker II. Nevertheless, the mistake resulted in an international uproar, including renewed charges that the U.S. was "carpet bombing" North Vietnam's capital. In reality, Linebacker II crews, often at their own peril, took extraordinary measures to avoid civilian casualties. By the end of the 11-day campaign, even official North Vietnamese figures put civilian dead in Hanoi at just 1,318, with another 306 in Haiphong. When contrasted against actual U.S. carpet bombing raids during World War II, those numbers pale in comparison. For example, during one campaign against the German city of Hamburg in 1944, American flyers killed more than 40,000 civilians in just nine days.[30]

Despite the last-minute change in tactics, two more BUFFs had been lost. Meyer sent word to Eighth Air Force headquarters on Guam. "Events of the past four days produced significant B-52 losses which obviously are not acceptable on a continuing basis." He ordered staff to develop comprehensive plans that would include varying "B- 52 flight altitudes with the chaff corridor on ingress. Change release altitudes and the ingress/egress headings on a daily basis" among other requirements. The general wanted his action plan in place by 26 December.[31] Until the Eighth Air Force could get a handle on things, B-52s were to avoid Hanoi. Instead, planners shifted their focus to Haiphong, which had already been subject to significant tactical strikes since the campaign's start. U-Tapao would assume responsibility for Linebacker II missions in coming days. Meanwhile, Andersen crews were relegated to flying Arc Light strikes in the South while planners searched for solutions.[32]

On 22 December, F-111s conducted night attacks on 10 Hanoi-area targets, hitting airfields, transshipment hubs, radar and communication assets, port facilities, and more. Antiaircraft gunners managed to shoot down one Aardvark as it was attacking a port facility on the Red River. The two-man crew of the 429th TFS out of Takhli successfully punched out but were taken prisoner.[33] A single wave of 30 U-Tapao BUFFs then bore down on Haiphong, tactics and targets reflecting SAC's updated approach. Rather than come in over land, the B-52s looped around and approached from the Tonkin Gulf. SAM suppressors—this time Navy A-6 Iron Hands—went in first, striking nearly a dozen SAM sites, while Navy A-3 Skywarriors and Air Force EB-66s stood off shore to jam enemy radar and communications. As usual, chaff Phantoms followed but this time employed a new technique. Rather than spread the uneven chaff clouds thought to confuse enemy crews during Linebacker I, the F-4s instead dispensed extra-thick layers right over the Fan Song units to prevent so-called "burn through." Generally, the closer an aircraft got to a land-based radar unit, the easier it was for the more-powerful ground station to punch through its ECM jammers. Planners hoped that blanketing enemy acquisition radar with chaff would effectively smother them.[34]

As the B-52s approached, gone was the long, continuous train of days past, Instead, the BUFFs came in at varying altitudes, distances, and time intervals on three separate tracks. As they closed on the city, the tracks spilt once again into six smaller packets. None headed directly for their intended targets—Haiphong's main POL storage facility and its railyard complex. Instead, some packets feinted west toward Hanoi, while others took different tracks. At just 30 miles from their intended release point, the BUFFs changed tracks again, reassembling just off shore and zeroing on their targets. Both objectives sustained massive damage, with secondary explosions, raging fires, and billowing black smoke testaments to the bomber crews' success. Undoubtedly taken by surprise, enemy gunners managed to fire off just 43 SA-2s, a low for the campaign. More significantly, no BUFFs were downed or even significantly damaged. It was welcome news for SAC and marked a turning point in the campaign.[35]

Meanwhile, daylight tactical strikes raged on. Bad weather on 23 December forced the cancellation of a LGB strike against rail bridges north of Hanoi. Once more, Air Force F-4 Pathfinders stepped up to maintain mission tempo. The LORAN-equipped Phantoms guided 32 F-4s against radio and communication facilities in the capital. Another 27 all-weather A-7Ds pummeled the Hoa Lac airfield with 500-pound MK-82s. The Navy kept up its daytime pressure as well, with Intruders and Corsairs hammering enemy air defenses, railyards, storage facilities, power plants, and shipyards in Haiphong. Antiaircraft fire was thick, but few SA-2s were launched, giving further credence to intelligence reports indicating a growing missile shortage. The communist high command, it seemed likely, was saving its SAM stocks in hopes of downing more of the big Stratofortresses.[36]

That night, the B-52s returned but with yet another twist. Rather than striking either Hanoi or Haiphong, a 30-plane wave surprised the enemy by pummeling rail facilities north of Haiphong and just 18 miles from the Chinese border. Eighteen Deltas from U- Tapao and another 12 Ds from Andersen—marking the 43rd SW's first participation in three days—dumped 800 tons of high explosives on the Lang Dang yard. Six of the Andersen birds diverted to blast three nearby SAM sites, as well. At the same time, F-111s hit the MiG bases at Kep, Phuc Yen, and Yen Bai to discourage enemy fighter activity. A few MiG - 21s managed to get airborne to pursue the departing bombers, firing four Atoll air-to-air missiles, all of which missed. None of the four or five SAMs launched hit their marks either. Although a F-4J flying photo reconnaissance escort was shot down by AAA fire and an EB-66 was lost to engine failure, 23 December marked the second day in a row that no B-52s had been lost.[37]

Christmas Eve saw 30 U-Tapao Deltas loop around Hanoi's lethal SAM zones to strike the Thai Nguyen and Key rail complexes 40 miles north and northeast of the capital. Enemy defenders fired about 20 SA-2s, but just one bomber sustained minor damage. MiG fighters again rose to target the BUFFs but did not score a hit. In fact, it was the other way around. Airman 1st Class Albert Moore was riding shotgun for the 307th's *Ruby 03* when his radar scope picked up a blip at eight miles. In a flash, the bandit had closed to within 4,000 yards. Moore called for evasive action, chaff, and flares. At 2,000 yards, the gunner opened up with his quad-50, burning through 800 rounds in three controlled bursts until the blip suddenly flared and then disappeared. For just the second time in the war a B-52D tail gunner had bagged a MiG-21. Moore would later be awarded the Silver Star for the action, one that inspired both regret and satisfaction for the airman. "There was a guy in that MIG," he would later say. "I'm sure he would have wanted to fly home too. But it was a case of him or my crew. I'm glad it turned out the way it did."[38]

At midnight began a 36-hour bombing hiatus as a goodwill gesture for the Christmas holiday. For the men of U-Tapao, the 36-hour standdown would be a welcome respite, indeed. Although Andersen's 43rd SW had sent 22 Delta crews to Thailand to help shoulder the burden, 307th and 310th ground and air crews had been grinding around the clock to maintain Linebacker II mission tempo. The strain on men, machines, and material had reached its breaking point. The respite was to be short-lived. Planners at U-Tapao and Andersen were warned to prepare "maximum effort" strikes beginning the night of the 26th. The mission would be the "most ambitious to date," an all-out push to end the war on terms acceptable to the United States.[39]

Daylight tactical strikes resumed on 26 December. Bad weather again precluded the use of LGB munitions, so 32 Air Force A-7Ds and 16 LORAN-guided Phantoms struck Hanoi's main electrical transformer station, damaging the facility and temporarily compromising the city's electrical grid. SAM launches were nearly non-existent as the North's missile stockpile dwindled.[40] At the same time, four carrier wings off the *Sarasota, America, Ranger,* and *Enterprise* resumed attacks on Haiphong with gusto.[41]

But it was that night that the heavy hammer of Nixon's "maximum effort" would truly fall. A force of 120 Stratofortresses, including 45 B-52Gs and 33 Ds from Andersen, and another 42 U-Tapao Deltas, set upon North Vietnam's industrial heartland in a single mass assault. Ten waves, each with its own target, would strike objectives in the Hanoi-Haiphong complex. The weight of the attacks would fall on Hanoi, with 90 BUFFs dedicated to pounding rail yards, storage facilities, and POL storage depots in and around the capital. The other 30 B-52s would strike Haiphong's main electrical station and railyard. Although routes, altitudes, and timing intervals between cells were varied in accordance with hard-won lessons, all bombers would have the same initial time on target. Strikes were to be completed within 15 minutes to deliver "maximum impact on the enemy's defense network…[and] oversaturate his command and control system."[42] Support aircraft topped 75 planes for the first time. Twenty-three Air Force F-4s dumped vast corridors of chaff over target areas, while additional F- 4Es armed with cluster munitions targeted antiaircraft defenses. Dozens more F-4s from both services flew combat air patrol, while Navy Iron Hands worked over SAM and AAA sites in and around Haiphong. Task Force 77 also sent in Marine EA-6A Prowlers from Da Nang to help jam enemy radar. In a new high for the campaign, Air Force and Navy would fly 113 support sorties over the 24-hour period.[43]

Once again, the venerable F-111s preceded the big bombers, conducting whispering death attacks on the MiG bases at Kep, Hoa Lac, Phuc Yen, and Yen Bai. At just after 2230, the BUFFs arrived on target. Nixon had hoped his Christmas truce would encourage Hanoi to resume negotiations. Instead, the communist high command had taken the opportunity to reset and restock its air defenses. Soon the sky filled with AAA flak and the most SA-2 launches in days. "The radio was completely saturated with SAM calls and MiG warnings," recalled 43rd

SW commander Col. James McCarthy, who led his unit's strikes that night. "At 26 SAMs, I stopped counting. At bombs away, it looked like we were right in the middle of fireworks factory that was in the process of blowing up." And still the BUFFs came on. Regardless of other tactical changes, the need to stabilize bombing computers just prior to release still precluded evasive action. Only seconds from bomb release, the crew of the 307th's *Ebony 02* watched as an SA-2 closed for the kill. "The copilot calmly announced the impending impact to the crew over the interphone," said McCarthy. "The aircraft dropped its bombs on target and was hit moments later. That's what I call 'guts football.'" Flames engulfed the plane in mid-air before it finally crashed southwest of Hanoi. Miraculously, four crewmembers managed to eject and were taken prisoner. A second 307th Delta, *Ash 01,* was badly damaged by SAM fire during its run over the Kinh No rail complex. Red Crown vectored F-4s to keep enemy fighters off the stricken ship while Capt. James Turner coaxed the BUFF to safety in Thailand. But despite a heroic effort, *Ash 01* crashed just beyond the U-Tapao runway, killing Turner and three other airmen. Only the gunner and copilot survived.[44]

Amid the chaos, an errant B-52 bomb train laid waste to an 18-block stretch of Kham Thien Street in Hanoi, tragically killing some 200 civilians and injuring 257 others. It would become Linebacker II's worst single incident of collateral damage, sparking domestic and international outrage. Although much of the area had been evacuated prior to the renewed bombing, some residents had returned during the Christmas truce. Then a boy, Le Dinh Giat recalls hunkering in a shelter as the bombs fell. "The dark bunker shook violently," he said. "I was flipped upside down again and again." Once the explosions stopped, Giat dug his way out inch by inch "like a mole." On the surface, he discovered his street in ruins, bomb craters strewn with bricks and stones, the roofs of houses blown off. Giat lost most of his family that day, his father, mother, and sister killed. Just he and a younger sister survived.[45]

The next day, the politburo sent word to the White House that it was willing to resume the Paris talks. To prevent the impression that the Linebacker II raids had re-instilled an eagerness to negotiate, the politburo did not condition renewed talks on a bombing halt. Hanoi proposed 8 January to begin resolving "remaining questions." While Nixon welcomed the news, he wanted some form of resumption before the new 93rd Congress was sworn in during the first week of January. He pushed for lower-level discussions of technical issues to begin on 2 January. If Hanoi agreed, Nixon pledged to end all bombing north of the 20th parallel within 36 hours of the two sides publicly assenting to the arrangements. On 28 December, Hanoi sent word that it accepted. The next day, the president publicly announced that Deputy Assistant Secretary of State for East Asian and Pacific Affairs William H. Sullivan and North Vietnamese Deputy Foreign Minister Nguyen Co Thach would resume technical discussions on 2 January.[46]

In the meantime, Nixon had no intention of letting up. The night of 27 December brought more strikes on North Vietnam's capital region. Haiphong had seen the last of the B-52s. This time, 36 Ds from U-Tapao and Andersen targeted railyards and storage depots in Hanoi, with 21 of the more vulnerable Guam-based Golfs restricted to relatively safer attacks on the Lang Dang rail complex near the Chinese border. In addition to the usual slate of airfield targets, F-111s also swept in to hit several SAM batteries for the first time. This was part of the intensifying—yet increasingly frustrating—U.S. effort to decisively cripple Hanoi's air defenses. From the start, exceptionally poor weather had severely limited SAM site attrition, with just 12 hours of acceptable daylight bombing weather throughout the course of Linebacker II.[47]

Earlier in the day, H-Ks had pummeled an array of missile sites with radar-homing missiles and cluster munitions, while Navy Iron Hands blasted SAM emplacements east of the capital. The BUFFs joined the all-out offensive, as well. Of particular interest was a SAM site dubbed "Killer Site VN-549" southwest of Hanoi. The problem, however, was that the BUFFs were much more successful against area objectives than against pinpoint targets like SAM sites. VN-549 not only survived but launched a counter-salvo just moments after being bombed, adding the 307th's *Ash 02* to its growing list of bomber kills. Ironically, *Ash 02* had been targeting another SAM site when its egress route took it within range of VN-549's gunners. Pilot Capt. John Mize nursed the stricken and burning craft into Thai airspace before the crew ejected near Nakhon Phanom. All survived and were quickly rescued. Mize would be awarded the Air Force Cross for the feat, becoming the first and only SAC airman of the war to be so honored.[48]

Meanwhile, a 43rd SW BUFF, *Cobalt 02*, was caught in a barrage of SA-2s during its run over the Truang Quang railyard northeast of Hanoi, rupturing the big bomber's fuel tanks and setting it ablaze. Navigator 1st Lt. Ben Fryer was killed in the explosion. Despite suffering shrapnel wounds, four other airmen managed to eject and were taken prisoner. The ship's EWO, Maj. Allen Johnson, was declared MIA. The day had indeed proved costly. Aside from the lost BUFFs, a pair of F-4Es had also been shot down by MiG- 21s, and small arms fire had claimed an Air Force HH-52 Jolly Green rescue chopper.[49]

The next two nights would each see 60-plane attacks in the capital region. Again, SAM sites and missile storage depots received extra attention, as did the Lang Dang railyards near the Chinese border. The latter was an important nexus for overland imports, and U.S. planners were keen to cripple the hub to prevent the North from quickly resupplying once a ceasefire arrived. F-111s hit SAM sites at night, while Wild Weasels sustained the pressure during the day. Another 32 Air Force A-7Ds added extra muscle to the ongoing war on SAM capabilities. Finally, after days of combined B-52 and tactical strikes, the effort began to pay off. Crews reported an average of just 20 SA-2 launches on 28 and 29 December. The Weasels also reported a marked drop in enemy radar signals. Overall, antiaircraft fire fell dramatically across the board, with no U.S. aircraft lost to groundfire over the two days. Heavy damage to enemy SAM support and supply facilities at Trai Ca and Phuc Yen helped ensure that the North's capability would not recover anytime soon. One Navy RA-5C was shot down by a MiG-21 on 28 December, but it would mark the final U.S. aircraft lost during Operation Linebacker II.[50]

Meanwhile, U.S. fighter escorts scored a few last-minute victories. Two separate air-to-air clashes on 28 December saw an F- 4D with the 555th TFS dispatch a MiG-21, while an F-4J with VF- 142 off the *Enterprise* splashed another. These would mark the last MiG kills of the campaign…but not the war. That honor went to Lt. Vic Kovaleski and his RIO Jim "Wizzer" Wise on 12 January when the USS *Midway* duo claimed one final enemy fighter. Ironically, aircraft off the "Midway Magic" had scored the first two aerial victories of the war on 17 June 1965.[51]

At 0017 on 29 December, the last BUFF cells pulled off their targets over North Vietnam. Operation Linebacker II was over. Earlier that afternoon, the JCS had sent word to Eighth Air Force's Lt. Gen. Gerald Johnson that all bombing north of the 20th parallel would halt at 0700 Hanoi time. Air raids would continue below the parallel but at a much reduced tempo. From 18 to 29 December, 200 B-52s out of U-Tapao and Andersen had flown 729 bombing sorties into the teeth of North Vietnam's vaunted air defenses, delivering more than 15,000 tons of bombs on 34 targets, mostly in the Hanoi- Haiphong complex.

Estimates vary, but ground crews reportedly fired between 800 and 1,200 SA-2 missiles in response, downing 15 Stratofortresses, including nine Deltas and six Golfs. Despite the seemingly jarring numbers, however, this constituted just a 2.06- percent loss rate, well short of the 3 to 5 percent expected by planners. Another 10 B-52s were damaged, three severely. Thirty- three SAC aviators were killed or missing, and another 33 made POWs. Of the 31 BUFF airmen to reach Laotian or Thai airspace, 24 were rescued.[52]

Meanwhile, Air Force, Navy, and Marine aircrews flew more than 1,200 strike and support sorties. Tactical strike bombers added another 5,000 tons of ordnance to the tally, while escort and combat air patrol fighters downed four MiGs. B-52 tail gunners would claim a pair of MiG-21s, as well.[53] In return, the North Vietnamese mustered MiGs, launched SAMs, and raked the skies with an untold number of antiaircraft rounds to ward off the attackers. The Air Force lost two F-111As, two F-4Es, and an HH-53 Jolly Green CSAR chopper in the effort. An EB-66C was also lost to engine failure. Hostile fire claimed another six Navy and Marine aircraft, including two A-6s, two A-7s, an F-4J, and an RA-5C. In all, the strike and support effort cost the U.S. 10 airmen killed and eight taken prisoner. Another 11 were rescued.[54]

The bombing campaign had indeed been costly for the U.S., but its end had come not a moment too soon for the VWP politburo. North Vietnam's vaunted air defenses had been shattered. Six months of naval blockade, overland interdiction, and relentless air attack had depleted or destroyed its SA-2 stocks, while what remained of its once-potent fighter inventory had been gutted. At long last, North Vietnam was at the full mercy of American airpower. The rest of the country had suffered mightily, as well. Post-strike bomb damage assessment (BDA) revealed an economic, military, industrial, and transportation infrastructure in ruins. Nearly every aspect of the economy had been wrecked, from the production of coal, electricity, and fertilizer to machine tools, textiles, and more. During the 11 days of Linebacker II alone, some 1,600 military structures had been damaged or destroyed, rail transport substantially cut at 500 separate locations, nearly 400 cars of rolling stock eliminated, a quarter of the country's POL stocks wiped out, nearly every airfield above the 20th parallel wrecked, approximately 80 percent of electrical capacity destroyed, and untold numbers of stockpiles, repair, and maintenance facilities battered. Moreover, material imports, which had crept back up during Nixon's eight- week bombing pause, once again plummeted from their pre-Linebacker levels of 160,000 tons per month to just 30,000 tons by January.[55]

The U.S. air campaigns of 1972 had battered the politburo's dream of building a socialist economy in the North. That year, nearly a third of North Vietnam's annual budget for economic development had been absorbed by ongoing efforts to repair transportation and communication lines. That was nearly three times what had been allocated for 1961-1964, and roughly the same amount devoted during the heyday of Operation Rolling Thunder. Meanwhile, tens of thousands of laborers were forced to divert from economic and military activities to repair and otherwise mitigate the destruction. The air war had virtually undone all of the progress made in the three and half years since Johnson halted the bombing in November 1968. By the end of 1972, one French foreign affairs adviser who traveled frequently to North Vietnam concluded that Hanoi's prospects for continuing the war seemed unworkable due to "the almost complete destruction in the North." None other than Le Duan was forced to admit later that the bombing had "completely obliterated our economic foundation."[56]

The air campaign had been keenly felt by forces fighting in the South, as well. By late summer 1972, less than 20 percent of pre-

Linebacker supply levels were reaching North Vietnamese units at the front.[57] "Our cadres and men were fatigued, we had not had time to make up for our losses, all units were in disarray, there was a lack of manpower, and there were shortages of food and ammunition," recalled Gen. Tran Van Tra, a commander in southern South Vietnam. According to the VWP's official history of the war, despite "massive assistance from the North, the effort to maintain our troop strength and supply levels had not kept pace with the requirements of the battlefield." NVA battalions in the South had been reduced to an average of just 200 men, less than half of their authorized strength.[58] Beyond those killed and seriously wounded, many units had been stripped of their most experienced troops and cadres so they could return North to either help boost economic output or fight to defend the homeland. Defections also increased, as many who had long fought under difficult conditions in the South came to view continued hostilities as futile. Such hardships convinced many communist leaders to push for peace—however imperfect—in order to recuperate, consolidate, and reorganize for the final push to conquer South Vietnam. "We needed time to build up the country as well as our forces," recalled Truong Chinh, chairman of North Vietnam's Standing Committee of the National Assembly. Hanoi's communist patrons also encouraged a return to diplomacy. As Chinese foreign minister Zhou Enlai counseled, "Let the Americans leave as quickly as possible. In half a year or one year the situation will change."[59]

Nixon understood that his adversaries in Hanoi were reeling. Haig and senior military commanders urged him to press the advantage to secure a more advantageous peace agreement. Such reasoning is epitomized in the thoughts of British Southeast Asia expert Sir Robert Thompson, who had helped defeat the 1948-1960 communist insurgency in Malaysia. "In my view, on December 30, 1972, after eleven days of those B-52 attacks on the Hanoi area, you had won the war," Thompson was quoted in 1977. "It was over! They had fired 1,242 SAMs; they had none left, and what would come in overland from China would be a mere trickle. They and their whole rear base at that point were at your mercy. They would have taken any terms. And that is why, of course, you actually got a peace agreement in January, which you had not been able to get in October."[60]

But Nixon chose to call off the dogs. The president was laboring under his own set of constraints, including congressional and domestic pressure that threatened to boil over at any moment. Even before the new Congress took office, the House Democratic caucus fired yet another shot across the bow on 2 January, voting 154-75 to completely cut off funds for military activity in Indochina once the American withdrawal was complete and the POWs came home. The Senate Democratic caucus followed suit two days later by a vote of 36 to 12. This from a body that had seen just 19 senators in favor of the Linebacker II campaign and 45 lawmakers opposed, according to a survey conducted by *Congressional Quarterly.* Iowa Sen. Harold Hughes (D), likened the so-called "Christmas bombings" to that of Hiroshima and Nagasaki, declaring, "It is unbelievable savagery that we have unleashed in this holy season." Sen. Mike Mansfield, another Democratic critic, called the campaign a "Stone Age strategy" and "a raw power play with human lives, American and others…it is abhorrent." The press maintained a steady drumbeat of criticism, as well. *New York Times* columnist Anthony Lewis accused Nixon of behaving like a "maddened tyrant," while colleague James Reston condemned the president for conducting "war by tantrum." *The Washington Post* charged Nixon's 11th-hour bombing campaign with being "so ruthless and so difficult to fathom politically as to cause millions of Americans to cringe in shame and to wonder at their President's very sanity." Columnist Joseph Kraft, a longtime tormentor of the president, intoned that the bombing was "senseless terror which stains the good name of America."[61]

One group that had a decidedly different take was the American POWs held in the infamous Hanoi Hilton. As the distant rumble of B-52 bomb trains shook the earth, the morale boost was incalculable. For the first time, they saw fear in the eyes of North Vietnamese guards who had long terrorized them. And they also realized that that their country had not abandoned them. Air Force Col. Robinson "Robbie" Risner, who had suffered nearly eight years of torment at the hands of his captors, recalls the first night from his vantage point in the Hanoi Hilton. "There was never such joy seen in our camp before," he recalled. "People jumping up and down, putting our arms around each other, tears running down our faces. We knew they were B-52s and that President Nixon was keeping his word and the Communists were getting the message." Navy Capt. Howard Rutledge, more than seven years a POW, tells the story of one guard, nicknamed "Parrot" by the Americans. Steeped in his country's propaganda, Parrot was terrified that U.S. flyers would "carpet bomb" the prison, indiscriminately killing Vietnamese and American alike. After fleeing the Hilton to find a "safe" place, he slunk back in among the prisoners, wild-eyed and shellshocked by the destruction he had seen. Taking pity, the Americans told him, "Don't worry. Stay with us. We'll protect you," Rutledge remembers. Col. John Flynn, an Air Force flier imprisoned since October 1967, was elated. "When I heard the B-52 bombs going off," he says, "I sent a message to our people. Pack your bags. I don't know when we're going home—but we're going home."[62]

10

PAX INFIDUS

January 1973

On 8 January, Kissinger and Tho reconvened for the 23rd private session, again at Gif-sur-Yvette. After the initial "frostiness" thawed, the two delegations set about making steady progress toward a final agreement. The next day, the men finalized language on the DMZ, a major sticking point for both Hanoi and Saigon. While retaining language that described the line as "provisional and not a political or territorial boundary," the wording did call on North and South to negotiate civilian movement across the boundary, a tacit acknowledgement that, pending "peaceful" unification, the two remained distinct political entities. Most important from the U.S. perspective, however, was Tho's return to his earlier concession that the release of American POWs would not be linked to the status of detainees held by Saigon. Instead, the GVN and PRG would "do their utmost" to resolve the issue within 90 days "in a spirit of national reconciliation and concord." Kissinger's optimism for a quick settlement returned.[1]

In his report to Washington, however, the envoy warned against replaying the giddiness that had afflicted the administration in October. Secrecy, more than ever, was paramount. "I cannot overemphasize the absolute necessity that this information be confined to the President alone. There must not be the slightest hint of the present status to the bureaucracy, Cabinet members, the Congress, or anyone else. If a wave of euphoria begins in Washington, the North Vietnamese are apt to revert to their natural beastliness, and the South Vietnamese will do their best to sabotage our progress." Kissinger added that statements by Tho had convinced him that Linebacker II had prompted Hanoi's renewed enthusiasm to settle. It was therefore imperative that Nixon continue to evince firmness in the days ahead. "The slightest hint of eagerness could prove suicidal," he cautioned.[2]

Further movement had occurred by 11 January. In a series of unwritten, "secret understandings" between Tho and Kissinger, the former consented to shorten the interval between ceasefires in Vietnam and Laos from 30 to 15 days, and both sides agreed that no foreign forces could use either Laos or Cambodia as a staging ground for attacks. Kissinger asked for "iron clad guarantees" on the return of American POWs in both countries. Tho consented on Laos but, citing his government's deteriorating relationship with the Khmer Rouge communist insurgency, could not guarantee either a ceasefire or prisoner return for Cambodia. Nevertheless, Kissinger reciprocated by pledging Washington's assistance in getting Saigon to release its prisoners in the South. Meanwhile, Kissinger got no concession on the PRG and its implied status as both a legitimate governing authority in South Vietnam and lawful signatory to the final settlement. The two compromised by simply referring to the "four parties" in the agreement. The U.S., DRV, GVN, and PRG would sign this document, while only the U.S. and DRV would discreetly sign a second that named the communist shadow government. Much more ominous, however, was that there again had been no concrete movement on withdrawing NVA troops from South Vietnamese territory. Instead, the two South Vietnamese parties, "as soon as possible" and in a spirit of "equality and mutual respect," were to negotiate steps for demobilizing and reducing the effectiveness of all "Vietnamese" forces in South

Vietnam.[3]

Thieu had repeatedly stressed, including in his most recent letter to Nixon on 7 January, that both legitimizing the PRG and the presence of NVA troops were "life or death issues for all the people of South Vietnam" and would likely prove fatal if not resolved. Hanoi, he wrote, "has not abandoned its objectives over South Viet Nam, and makes no secret about it." Regardless of what the U.S. sacrificed at the bargaining table, Thieu continued, "Any concessions we shall make to the Communists will be theirs forever, while they consider any compromises they would make as only temporary." The South Vietnamese president concluded by reiterating his desire for peace "with honor and with justice, a peace which could justify all the sacrifices we have made in this long struggle for freedom."[4]

Nevertheless, by 13 January, Kissinger and Tho considered the 23-article "Agreement on Ending the War and Restoring the Peace in Viet-Nam" complete and settled on a signing and implementation schedule. First, all bombing and mining of North Vietnam would cease on 15 January, with Nixon announcing the end of American military operations in Vietnam four days later. Kissinger and Tho would then initial the agreement in Paris on 23 January and publicly announce the act the next day. On 27 January, the four parties, including the GVN and PRG, were to gather in Paris to sign the final agreement, with a ceasefire-in-place commencing immediately.[5]

The U.S. and its foreign allies were to withdraw all military forces, dismantle installations, and remove military, civilian, and technical advisers within 60 days. Release of American POWs throughout Indochina would coincide with that withdrawal. Additionally, Article 20b called on "foreign countries" to end all military activities in Laos and Cambodia, withdraw all forces and advisers, and refrain from reintroducing any military assets. The signatories would designate representatives for a Four-Party Joint Military Commission (FPJMC) to oversee the ceasefire and withdrawal during the first 60 days. Once complete, a Two-Party Joint Military Commission (TPJMC) composed of military personnel from both sides would ensure "joint action" between the two Vietnamese parties, including the ceasefire and other military provisions, areas controlled by each party, and the "modalities of stationing." Article 3c instructed the regular and irregular forces of both South Vietnamese parties to "stop all offensive activities against each other," including "all hostile acts, terrorism and reprisals." Further, Article 7 prohibited both from accepting the "introduction of troops, military advisers…and technical military personnel" into South Vietnam. Nor, Article 7 continued, should either party accept any "armaments, munitions, or war material." A basic piece-for- piece replacement of destroyed, damaged, or worn out equipment was authorized but only under TPJMC and international supervision.[6]

On the political front, the two parties were to form a National Council of National Reconciliation and Concord (NCNRC) to promote the agreement's implementation, ensure democratic liberties, and organize the internationally supervised "free and democratic" general elections called for in Article 9b. An International Commission of Control and Supervision (ICCS) composed of delegates from Hungary, Poland, Canada, and Indonesia would provide outside oversight, including arbitrating disagreements between the Vietnamese parties. Ultimately, the process and timing of reunification were to be negotiated "step by step through peaceful means on the basis of discussions and agreements between North and South Viet-Nam without coercion or annexation by either party and without foreign interference."[7]

The course now set, Nixon's "honorable peace" still demanded he make one last-ditch effort to gain Thieu's acceptance. In his report to Washington, Kissinger urged the president to take a firm hand with Thieu to prevent him from again scuttling the hard- brokered settlement. "I believe the only way to bring Thieu around will be to tell him flatly that you will proceed, with or without him. If he balks and we then initial, there will still be 3 to 4 days between initialing and signing for the pressures to build up. I have already told Le Duc Tho that we would have to discuss the situation in this eventuality. In any event, if we once again delay the initialing or reopen the negotiations, we would not only jeopardize but certainly lose everything that has been achieved." Nixon quickly endorsed his envoy's approach. "We must go ahead with the agreement with Hanoi regardless of whether Thieu goes along or not." If they could not secure Thieu's acquiescence, Nixon continued, he would announce on 22 January that "we had reached an agreement in principle with the North Vietnamese and call on Thieu to adhere to it." Because the relationship between Kissinger and Thieu had so deteriorated during the October controversy, Nixon again turned to Haig. But, the president warned, "I have already told Haig that he is to tell Thieu that we are not going to negotiate with him but rather that we will proceed and we are presenting this, in effect, on a take- it-or-leave-it basis."[8]

On 18 January, Haig and Ambassador Elsworth Bunker met with Thieu in Saigon and delivered a letter from Nixon. Prior to the meeting, Kissinger had cabled Bunker instructing him not to be drawn into further negotiations with the South Vietnamese president. Rather, Haig's summary of the agreement—and most importantly, Nixon's letter—should be left to speak for themselves. The president's letter began by reiterating his longtime commitment to South Vietnam and enumerating the "many grave domestic and international consequences" he had endured in support of it. Nixon listed aspects of the agreement he believed had been improved to South Vietnam's benefit. But, he warned, the time for negotiation had passed. "The text of the agreement, the method for signing, and the protocols are the best obtainable," he wrote. "They can no longer be changed." Continued resistance would only result in a "total cutoff of funds" by the U.S. Congress. "The key issue," Nixon continued, "is no longer particular nuances in the agreement but rather the post-war cooperation of our two countries and the need for continued U.S. support. If you refuse to join us, the responsibility for the consequences rests with the Government of Vietnam." Therefore, the president declared, he intended to move forward with or without Thieu. The president concluded by urging Thieu to "continue in peacetime the close partnership that has served us so well in war…to join together at last and protect our mutual interests through close cooperation and unity."[9]

Thieu responded with a letter of his own on 20 January. After repeating past objections— especially the continued presence of NVA troops in South Vietnam—the GVN president showed signs he was beginning to waiver. At the very least, he wrote, the agreement must ensure that Hanoi acknowledge and respect the DMZ as a permanent political boundary separating sovereign states and that ironclad provisions be included for the removal of North Vietnamese troops following the ceasefire. "I deeply believe that these proposals are most reasonable and are the very strict minimum indispensable to give the RVN a chance for survival, and therefore they deserve a last supreme effort vis-à-vis the Communist side."[10]

Sensing movement, Nixon replied immediately, listing several provisions he believed would assuage Thieu's final concerns, especially those over NVA troop presence in his country. The agreement, Nixon wrote, affirmed the independence and sovereignty of South Vietnam; provided for "reunification only by peaceful means…without coercion or annexation…[and] establishes the illegitimacy of any use or threat of force in the name of reunification; prohibited the "introduction of troops, advisers, and war material into South Vietnam from outside South Vietnam"; included a "principle of respect for the Demilitarized Zone and the Provisional Military Demarcation Line; and provided the opportunity for the GVN to negotiate the withdrawal of all communist forces "as soon as possible." On this last, Saigon retained a small measure of leverage in the more than 30,000 PRG operatives and sympathizers it still held. Finally, Nixon offered to issue a unilateral diplomatic note stating the United States' understanding of these issues.[11]

Left unsaid, of course, was that all of the foregoing would be completely contingent upon whether Hanoi decided to honor the agreement once the U.S. withdrew. Nixon envisioned a postwar reality modeled on the Korean example, where the threat of a U.S. military response—in this case, airpower stationed in Southeast Asia—would deter communist aggression. He also hoped it would make Saigon feel secure enough to honor its end of the bargain, too. Nixon had repeatedly assured Thieu that the U.S. would respond "massively" if the North significantly violated the agreement. And by all accounts, he meant it. Afterall, no settlement, regardless of its provisions, would alone thwart Hanoi's longtime goal of conquering the South. But whether a hostile Congress and a war-weary public would approve of renewed military action remained to be seen.[12] At any rate, Nixon closed with a final call to action. "This agreement, I assure you again, will represent the beginning of a new period of close collaboration and strong mutual support between the Republic of Vietnam and the United States. You and I will work together in peacetime to protect the independence and freedom of your country as we have done in war. If we close ranks now and proceed together, we will prevail." Still, the president concluded, "I must have your answer by 1200 Washington time, January 21, 1973."[13]

Under immense pressure, and with time running out, Thieu finally relented just hours before Nixon's deadline. But his return letter on 21 January reflected ongoing anxiety over the agreement. "Concerning the refusal by Hanoi to withdraw its troops from SVN at the conclusion of the cease-fire, I must say very frankly that I do not find that the collateral clauses you mentioned constitute an adequate remedy to this situation," he wrote. But the South Vietnamese president clearly understood his position. "However, for the sake of unity between our two Governments, and on the basis of your strong assurances for the continuation of aid and support to the GVN after the cease-fire, I would accept your schedule." In keeping with Nixon's recent offer, Thieu requested diplomatic statements affirming that the United States recognized the GVN as the sole legitimate governing authority in the Republic of Vietnam and that NVA troops had no right to occupy South Vietnamese territory.[14] Thieu then handed the letter to Bunker. "I have done my best," he said solemnly. "I have done all that I can do for my country." Later that evening, Thieu reportedly confided in his aide, Hoang Duc Nah. "The Americans really left me no choice—either sign or they will cut off my aid," he said. But Nixon had given him an "absolute guarantee to defend the country." Nha asked his president whether he could trust Nixon to keep his word. "He is a man of honor," Thieu replied. "I am going to trust him."[15]

Predictably, Nixon greeted the news with enthusiasm. In his responding letter of 22 January, the president pledged to inform Congress of Thieu's cooperation and declared his "great respect for the tenacity and courage with which you are defending the interests of your people in our common objective to preserve their freedom and independence." Per Nixon's instructions, Bunker then handed the GVN president a pair of diplomatic statements regarding the U.S. position on NVA forces in South Vietnam. In the first, Nixon averred that all communist forces in South Vietnam—whether Hanoi admitted to controlling them or not—must abide by the agreement, including the ceasefire-in-place, the prohibition on reinforcement and resupply, and the requirement that their reduction and demobilization be negotiated as soon as possible. Secondly, the president affirmed that no provision in the agreement permitted North Vietnam to maintain armed forces in South Vietnamese territory and that "the United States does not recognize any such right derived from any source." Now, wrote Nixon, the overwhelming concern was to "strengthen your government and people as we look toward implementation of the agreement" and stressed the need for "close cooperation and a confident approach" as the allies managed the endgame with Hanoi. "With your strong leadership and with continuing strong bonds between our countries, we will succeed in securing our mutual objectives."[16]

On 23 January, Kissinger and Tho and their respective delegations gathered in Paris to work out last-minute revisions and to initial "The Agreement on Ending the War and Restoring Peace in Viet-Nam." During the session, Tho agreed to honor an earlier U.S. request that American planes and crews land in Hanoi to retrieve POWs held in the North. Those held in South Vietnam, Laos, and Cambodia would be returned to Saigon or neutral sites in Southeast Asia. In return for the POWs' release schedule, Kissinger provided a timetable for final U.S. withdrawal. Moving to war reparations, Article 21 provided only that the U.S. would "contribute to healing the wounds of war and to postwar reconstruction" in the DRV but made no mention of what that would entail. Tho wanted a written guarantee that Washington would honor its verbal pledge to provide $3.25 billion for postwar reconstruction in exchange for North Vietnam's help in accounting for U.S. POWs in Laos. Kissinger said he could offer assurances on food and humanitarian supplies, but any monetary assistance would have to undergo normal congressional oversight. This was unsatisfactory to Tho, but he was apparently unwilling to scrap the deal at this late date.

The two agreed to revisit the topic during Kissinger's upcoming trip to Hanoi. With nothing further to discuss, the envoys initialed the agreement, bringing to a close more than three years of secret negotiations.[17]

The same day, Nixon announced the news in a nationally televised address. The president expressed hope that the agreement would "insure stable peace in Vietnam and contribute to the preservation of lasting peace in Indochina and Southeast Asia." Nixon reminded the audience of his repeated calls for an "honorable" end to America's war in Vietnam. Referencing his addresses on 25 January and 8 May 1972, the president affirmed that the agreement had at last accomplished that goal. Four years of secret and private negotiations had yielded an internationally supervised ceasefire, the prompt return of all American POWs and an accounting of those still missing, and the withdrawal of all U.S. forces from Vietnam. Moreover, the agreement provided the circumstances for an independent South Vietnam and the right of her people to determine their future free from external interference. To that end, Nixon continued, the settlement recognized the GVN as the legitimate governing authority in the country and preserved the right of the United States to continue to support its ally in peace as it had in war. To ensure a lasting peace, Nixon called on all parties to faithfully adhere to the agreement and pledged that the U.S. would do the same.[18]

The president then thanked the American people for their "steadfastness in supporting our insistence on peace with honor" rather than one that "would have betrayed our allies, that would have abandoned our prisoners of war, or that would have ended the war for us but would have continued the war for the 50 million people of Indochina." He expressed gratitude for the millions of Americans who had served in Vietnam and for those who had suffered and died so "that the people of South Vietnam might live in freedom and so that the world might live in peace." He also lauded the wives and families of POWs and those MIA for their refusal to give up. "When others called on us to settle on any terms, you had the courage to stand for the right kind of peace so that those who died and those who suffered would not have died and suffered in vain." Finally, Nixon called attention to Lyndon Johnson, a president who had suffered mightily under the strain of Vietnam. In a grim irony, Johnson had died just the day before the peace agreement had been initialed in Paris. After having "endured the vilification of those who sought to portray him as a man of war," Nixon intoned, "no one would have welcomed this peace more than he." For the sake of all who had died and those who yet live, Nixon concluded, "let us consecrate this moment by resolving together to make the peace we have achieved a peace that will last."[19]

On a cold and dreary Paris morning on 27 January, delegations from the U.S., RVN, DRV, and PRG gathered in an ornate ballroom of the Hotel Majestic. Signatories included U.S. Secretary of State William Rogers, Tran Van Lam, South Vietnam's foreign minister, North Vietnamese Minister of Foreign Affairs Nguyen Duy Trinh, and Nguyen Thi Binh, foreign minister for the PRG. There, amid crystal chandeliers and lush tapestries, the delegates gathered around the same large table that had witnessed so much acrimony over more than four years of public negotiations. In deference to Thieu, the four-party agreement omitted the PRG, instead noting only the "parties" that had negotiated the settlement. Later that day, Rogers and Trinh would sign a separate document that explicitly named the revolutionary government. Afterward, the gathered throng moved into an adjoining room to celebrate the occasion. After a somewhat chilly start, things soon livened as representatives from both sides clinked champaign glasses and engaged in cordial, even friendly banter. The only faux pas to disturb the generally convivial mood was when Poland's ambassador to North Vietnam congratulated Nguyen Duy Trinh on his country's *bonne victoire*—good victory—in front of GVN officials.[20]

However impolitic that may been in Paris, the sentiment was clearly shared back in Hanoi. "Our people in the North and in the South should be extremely proud and elated by this great victory of the Fatherland," proclaimed the VWP Central Committee. The signing not only marked the end of the "Anti-American Resistance" but heralded final victory over the South, as well. But much work remained. "The struggle of our people must continue to consolidate those victories and achieve still bigger new ones…[to] build a peaceful, unified, independent, democratic and strong Vietnam." The perspective merely reaffirmed what those in Saigon and even Washington had long understood—that no peace agreement would stand in the way of Hanoi's yearslong fight to unify the Vietnams under communist rule. Indeed, for Thieu and his supporters, the agreement had simply ushered in a new phase of the war. Forced to sign an agreement he abhorred, the South Vietnamese president could now only hope that Nixon would honor his pledge of continued U.S. support and retaliation when Hanoi violated the agreement. Whether the American president would be willing—or able— to honor that commitment remained to be seen.[21]

11

COMING HOME

February – April

At 0800 on 12 February, 1973, 27 U.S. prisoners of the Viet Cong waited in the red laterite dust of Loc Ninh, South Vietnam. Nearby, rows of American helicopters stood ready to ferry them to Tan Son Nhut Air Base 80 miles to the south. Meanwhile, U.S. C-130s were to fly hundreds of Viet Cong captives held by the ARVN from Bien Hoa to Loc Ninh. This was the plan. But there was already a problem. Some of the Viet Cong at Bien Hoa had balked at returning, and so the big aircraft sat idle. So too did the 19 U.S. servicemen and eight civilians, most of whom had suffered years of harsh captivity in the South. Brig. Gen. Stan McClellan, head of the American delegation onsite, vowed that nothing short of "doomsday" would prevent his team from seeing all U.S. POWs safely aboard the choppers. But communist representatives said no one was going anywhere until all Viet Cong POWs were enroute to Loc Ninh. An inauspicious beginning to a fledgling peace. And so the problem settled into impasse, and morning became afternoon. It was not until late that night that the Loc Ninh 27 would finally taste their freedom. But later is certainly better than never.[1]

Earlier that day, 700 miles to the north, a line of camouflaged busses groaned to a halt near the bomb-cragged hangar at Hanoi's Gia Lam Airport. These were the first 116 American POWs slated for repatriation from the North. Aboard the first bus, 20 U.S. servicemen sat in expectant silence. For the most part, they were the "Old Timers," those longest held in North Vietnam's notorious Hoa Lo prison – sardonically dubbed the "Hanoi Hilton" by its inmates. As a matter of code and honor, the men of the "4th Allied POW Wing," as they had named themselves, agreed that they would accept release only in the order of capture. In the front seat sat the Hilton's first "guest," 36-year-old Lt. (junior grade) Everett Alvarez, a Navy A-4 Skyhawk pilot who on 5 August 1964, became the first American shot down over North Vietnam. Beside him was Lt. Cmdr. Bob Shumaker, the second-longest held among those assembled. Suddenly, a great cheer went up among the men around them. Someone had spotted the silver and white visage of a C-141, tail emblazed with a big red cross, coming in for a landing. After eight and a half years of dashed hopes, Alvarez allowed himself to believe– truly believe–that this time it was real, that his day of deliverance had come round at last.[2]

Alvarez had been held by the Vietnamese longer than anyone except Army Special Forces Capt. Jim Thompson, captured by the Viet Cong after his 0-1 Bird Dog spotter plane was shot down near Quang Tri four months prior to Alvarez. Cruelly, Thompson's "Freedom Day" would have to wait. Although he had finally been brought to Hoa Lo the day after the peace agreement was signed, his release would not come until more than a month after this initial group. Kept separate from the Hilton long timers, Thompson and others later speculated that the North Vietnamese had held him longer hoping that an improved diet would mask years of starvation and torture. And there were many others. Like Alvarez, Shumaker, and Thompson, hundreds of other American prisoners of war had suffered mightily at the hands of their captors. But now, as the big C- 141— soon to be affectionately dubbed the "Hanoi Taxi" – rolled to a stop 50 yards away, the trials and tribulations of those years would soon be at an end.[3]

§

The moment was the culmination a years-long effort involving thousands of personnel from the departments of State, Defense, and the military services. As U.S involvement in Vietnam deepened in 1966, officials began preparing for the eventual repatriation of the accompanying surge of American POWs. Over the next few years, one of the most comprehensive and detailed operations of the war began to take shape. Planners worked through nearly every detail of repatriation, from transportation, logistics, and personnel to every conceivable need and want of returning POWs. The idea was simple: no effort would be spared in ensuring the returnees' physical, mental, and material wellbeing. For years, the burgeoning plan had been known by the rather prosaic moniker "Operation Egress Recap." But on 8 January 1973, outgoing Defense Secretary Melvin Laird—who had waged a personal crusade on the POWs' behalf since entering office in 1969—insisted that the operation have a name more befitting its driving ethos. To Laird, the "long-awaited repatriation of Americans captured in Southeast Asia deserved a more humane title." He ordered the name be changed to "Homecoming."[4]

According to the peace agreement, American POWs were to be released in increments proportional to withdrawals of U.S. forces from Vietnam. Operation Homecoming directed each increment be repatriated in three phases. Phase I saw the return of POWs to U.S. control via flights from either Hanoi's Gia Lam Airport, or from agreed-upon handover locations throughout Southeast Asia. The destination was Clark Air Base in the Philippines. Clark had been chosen as the Joint Homecoming Reception Center because of its proximity to Vietnam and its sprawling medical facilities. The stay at Clark was to be brief, ideally no longer than 72 hours. The goal was to get the men home as quickly and comfortably as possible. Each former POW was matched with a military service escort chosen for his similarity in age, rank, and personal interests. The escort was to build rapport with his man and help him navigate the initial medical exams, intelligence debriefings, uniform fittings, calls home to family, and so on. Escorts were to see their returnee's every need, perhaps acting as a dining companion one moment, arranging an after-hours trip to the base exchange the next. To be sure, many returnees had accumulated a healthy stash of backpay and were eager to spend it. Stereo gear and cameras were hot items, as were wardrobe updates to help transition to life in 1970s America. Escorts even helped arrange visits to children at nearby schools— heartwarming and therapeutic excursions that quickly became popular with many of the men.[5]

Once returnees were deemed fit for trans-Pacific travel, they and their escorts embarked on Phase II flights, usually to Travis Air Base in California. From there, Phase III flights delivered the men to some 31 military hospitals throughout CONUS, often nearest their homes and families. There, returnees underwent comprehensive medical and psychological treatment, as well as thorough intelligence debriefing on remaining POWs and those listed as missing in action. Phase III also helped returnees look to the future. Each was advised on financial planning and received public affairs guidance in case he wanted to grant media interviews or otherwise publish his experiences. There was career counseling, as well. And the message was clear: The sky's the limit. The U.S. government would do whatever possible to help returnees achieve their ambitions. Some went on to become doctors, others went to law school. Some returned to the cockpit, both as pilots and teachers, including as instructors for the Navy's Fighter Weapons School. Still others happily left the service to begin new lives as entrepreneurs, motivational speakers, authors, clergymen, and more.[6]

§

Back on the Gia Lam tarmac, Alvarez and the others climbed off the busses and formed a smart column, two abreast. Nearby, Air Force Col. James Dennett, head of the 18-man Reception Support Team and top U.S. negotiator on the ground, worked with his North Vietnamese counterpart, Lt. Col. Nguyen Phuong, to finalize the turnover. Dennett and his team had flown in hours earlier on an HC-130 Hercules from PACAF's 374th Tactical Airlift Wing to facilitate the transfer. Accompanying Dennett's team were 16 other personnel, including a flight surgeon, nurses, medical technicians, translators, photographers and public affairs specialists, and an airlift control crew. Also on board was a specially outfitted Jeep with an AN/MRC 108 mobile radio system which, in conjunction with another HC-130 circling off the coast, allowed for near real-time communication between the RST at Gia Lam and Clark Air Base, CINCPAC headquarters in Hawaii, and even the National Military Command Center in Washington D.C. The orbiting Hercules could also provide air rescue coverage in case any of the Phase I flights ran into trouble. Air Force Capt. Kenneth Green, commander of Clark's Aeromedical Evacuation Management Branch, had also come along for the inaugural flight. Because information on the health and overall condition of POWs was almost non-existent even at this late stage, Green wanted a firsthand look. He would use what he learned to better plan for the medical equipment and specialists needed to safely bring home future returnees. Finally, two civilians accompanied the recovery team: Dr. Roger Shields, 33, chair of the Defense Department's POW/MIA Task Force, and 38-year-old Frank A. Sieverts, the leading State Department official specializing in POW affairs. Through their energy, close working relationship, and dedication, both men were instrumental in shaping American POW and MIA policy for Operation Homecoming and beyond.[7]

The final signal came at last. With Alvarez and Shumaker at the lead, the first column marched in good order toward the airport terminal. Gone were the ragged and filthy POW pajamas the men had worn for years. Instead, the returnees were now clothed in the identical, light-colored zippered jackets and dark trousers the North Vietnamese had provided only days earlier. Many speculated that this, along with the improved diet and treatment in recent months, was their captors' attempt to put the best face on years of cruelty and abuse. In front of the terminal were American and North Vietnamese officials gathered at a white-cloth covered table shaded by a parachute canopy. Beyond, throngs of North Vietnamese military and civilians looked on. As a communist official called his name, each returnee was to step forward. To their fury, the men saw that the roll-caller was none other than "the Rabbit" himself, a particularly ruthless torturer nicknamed for his prominent ears and overbite. Dennett, initially unaware of the Rabbit's notorious history, moved to bar him from future repatriation ceremonies. But, as he later recalled, "I was informed by our men that it would make no difference whatever. The men were thoroughly disciplined, and nothing would happen to disrupt the proceedings."[8]

Alvarez stepped forward, saluted, and shook the hand of the receiving officer, Air Force Col. Al Lynn. A sergeant from the HC- 130 then took the Navy flyer gently by the arm. "C'mon, sir," he said. "We're gonna take you home." Alvarez fought back a sob as the American led him to the Hanoi Taxi. While the sergeant's move had been completely spontaneous, the practice of HC-130 crewmembers personally escorting returnees would become standard for all Phase I flights out of Hanoi. A gaggle of international reporters called after Alvarez for comment, but he and the others had been ordered by their senior ranking officers to speak only to U.S. representatives. An errant comment even now might endanger those still awaiting release. Every man moved forward after hearing his name, each dealing with the overwhelming emotions in his own way. Though held captive under terrible conditions for years, many limping from broken bones that had never properly healed, the men displayed a poise and military bearing that awed those who had come to take them home. Finally, three litter patients, too sick and hobbled to walk, were brought up by North Vietnamese bearers. At the midway point, American crewmembers came forward to carry their brothers to the waiting C-141.[9]

The massive plane had not been not the first choice for Homecoming flights. The Aeromedical Command argued that its C-9 Nightingales, well-appointed and purpose-built for medical transport, were ideal for the mission. But planners finally settled on the Military Airlift Command's C-141 for several reasons. First was the big cargo jet's 3,000-mile flight range—about 1,000 more than the Nightingale—making the Starlifter well-suited for both Phase I and II flights. Another was the 141's cavernous cargo bay. This allowed crews to configure each aircraft as needed, optimizing for unique passenger, crew, or equipment needs. The Starlifters were outfitted with both seats and litters to accommodate a returnee's preference or need. While the planes could carry many times the number, Homecoming officials limited each flight to no more than 40 returnees to maximize space and comfort. Standard operating procedure called for both primary and backup aircraft for each Phase I increment. For example, three C-141s had flown in to pick up this first group of 116 returnees, with a fourth orbiting just below the DMZ in case of trouble. Over Homecoming's duration—12 February through 4 April—C-141s would ferry some 567 returnees in 17 Phase I flights out of North Vietnam, while making 38 Phase II flights from Clark to Travis. MAC's 141s would ultimately bring home 591 former POWs. Meanwhile, the C-9 fleet was tapped to handle all Phase I flights from areas outside North Vietnam. The

Nightingales would conduct four flights, picking up the first 27 former captives from Tan Son Nhut and another three released by China to officials in Hong Kong. C-141s and C-9s shared Phase III duty.[10]

Back on the ground, Alvarez ascended the 141's ramp and spied something he had not seen in nearly nine years—an American woman. He stood awestruck gazing at the "delicate apparition" before him, a beautiful blonde flight nurse, early 30s, in a form- fitting uniform. "Every slight movement she made was divine," he would later recall.[11] While not aboard Alvarez's plane, flight nurse Lt. Mickey Mantel, then 25, says reactions like Alvarez's were common—and welcome—during her four Phase I and II flights. "They just loved talking to us," she says. "It wasn't like they were trying to pick us up or flirt. It just felt good for them to talk to another American— and a female. And we were so glad to be there for them."[12]

Nurses and medical techs escorted each returnee to his seat. The crew had brought aboard copies of *Stars and Stripes* and even a few Playboy magazines, along with aftershave, electric razors, and cloth hand towels—anything to make the flight more enjoyable. But on the food front, the men were to be disappointed. Homecoming dieticians were deeply concerned over returnees' digestive state after years of starvation and intestinal parasites. Instead, they prescribed "Sustacal," a bland high-protein drink. Capt. Kenneth Green, the Aeromedical Evacuation officer who had come along to assess the men's condition, later lamented the decision. "The first thing the men wanted when they got on board was something to eat. It was rather embarrassing to say all we have for you is Sustacal." Luckily, the onboard flight surgeon allowed nurses to give the men apples, chocolates, even ice-cold Cokes to tide them over. But Homecoming dieticians never changed the menu for subsequent flights, so medical crews resorted to smuggling aboard chocolate cake, salami, cheese, crackers, and so on.[13]

Bland diet or no, the men were ecstatic as the engines revved for takeoff. Alvarez, so close at last, prayed silently that there would be no breakdown, no last-minute disaster to snatch away hope. As the C-141 rumbled over the rough runway, pilot Major James Marrott poured on the power. All at once, the Starlifter's wheels pulled free of the earth, and a tremendous roar went up as the men cheered and laughed, backslapped and wept. After years of cruelty, mental and spiritual anguish, they were finally going home. But what world awaited? Their captors had fed them a steady diet of communist propaganda and news of antiwar protests at home. When Jane Fonda infamously "manned" her antiaircraft gun in July, guards gleefully shared the news. Had their country turned on them, the men wondered? The answer was not long in coming.[14]

Each man passed the four-hour flight to Clark Airbase in his own way. Some laughed and talked among themselves and with crewmembers. Many posed for pictures and autographed copies of *Stars and Stripes*. Others kept their own quiet counsel, each man dealing with the enormity of what he had experienced and what was to come. Soon, another raucous roar of hoots and howls exploded as the Hanoi Taxi's wheels touched down at Clark. Alvarez later wrote that during the flight it was agreed that he, as the longest held, would deplane first and say a few words. At some point, however, Alvarez says that Navy Lt. Cmdr. Jeremiah Denton, the senior ranking officer aboard, informed the junior officer that it would be he, not Alvarez, to deplane first and speak.[15]

When the moment came, Denton descended the ramp outside the C-141's jump door and down the red carpet. Ironically, given Homecoming's otherwise detailed planning, the carpet had been procured only at the last minute from Manila's Intercontinental Hotel. A thunderous cheer went up from the thousands who had come to welcome the men home. Air Force Lt. Col. "Robbie" Risner, a legend among the POWs for his years of bravery, leadership, and inspiration, was the SRO on the second flight that day. He and his men were overwhelmed by the crowd's response. "The sincerity and feelings in the welcome were beyond anything we had imagined," he later wrote. "Some were crying, many waving flags; they were just like our family."[16]

Denton saluted and then shook hands with CINCPAC Chief Adm. Noel Gayler and 13th Air Force Commander Lt. Gen. William Moore. He then stepped to the microphone, TV cameras carrying his words to the world. "We are honored to have had the opportunity to serve our country under difficult circumstances," he said. "We are profoundly grateful to our commander-in-chief and to our nation for this day. God bless America!" The crowd roared its approval. Similar scenes would play out over the next two months. At every stop of the returnees' long journey home, whether midday or midnight, in snow or freezing rain, the response had been the same—massive crowds and an outpouring of love, welcome, and gratitude for those who had sacrificed so much. Had their country turned on them? The men had gotten their answer in spades.[17]

As for Alvarez, he would indeed get his chance to speak. As he deplaned at Travis and walked his own stretch of carpet—this one red, white, and blue—he was welcomed by Maj. Gen. John Gonge, commander of the 22nd Air Force, along with thousands who had come to cheer his safe homecoming. After nearly nine years in brutal captivity, Alvarez addressed the crowd and the millions watching around the world. He spoke of faith, hope, and dreams through the long dark years. "We have come home," he said, finally. "God bless the president and God bless you Mr. and Mrs. America. You did not forget us."[18]

On 29 March, Nixon announced in a nationally televised address the end of America's long and divisive struggle in Vietnam. "For the first time in 12 years, no American military forces are in Vietnam, and all our POWs are on their way home," he said. "The 17 million people of South Vietnam have the right to choose their own government without outside interference, and because of our program of Vietnamization, they have the strength to defend that right. We have prevented the imposition of a Communist government by force on South Vietnam." Still, the president lamented the many problems in implementing the agreement, including gaining a full account of "all missing in action in Indochina, the provisions with regard to Laos and Cambodia, the provisions prohibiting infiltration from North Vietnam into South Vietnam have not been complied with." Nixon called on Hanoi to honor the agreement, warning that VWP leadership "should have no doubt as to the consequences if they fail to comply." He reminded the audience of the many difficult choices he had made concerning the war, especially his decision to renew air attacks on 18 December to break the deadlock in Paris. It was, he intoned, the "hardest decision I have made as President." He again thanked the "great majority" of Americans who had stood by him "against those who advocated peace at any price—even if the price would have been defeat and humiliation for the United States." Yet with the support of the American people, he continued, the nation had won an "honorable peace in Vietnam." America's longest war was over. Over nearly two decades, 2.7 million Americans served in Vietnam. More than 58,000 gave their last full measure on those distant shores. Another 150,000 sustained serious wounds. It was at times a bitterly divisive war. And yet, like each generation before them, America's Vietnam veterans answered their nation's call. How would their sacrifices be honored?[19]

Epilogue

REQUIEM

1973 – 1975

By the start of 1972, the United States was on its way out of Vietnam. Nixon had drawn down U.S. troop strength to levels not seen since early 1965. And more scheduled withdrawals were on the way. Meanwhile, the president's Vietnamization program was proceeding apace, with the South Vietnamese military better equipped and more effective than at any point in the war. All that remained was to "honorably" disengage from Vietnam. At its most basic, this meant reaching an agreement that secured the release of American POWs while enabling South Vietnam to survive as a sovereign nation-state.

But by summer 1971, the VWP was no longer interested in a negotiated settlement. While Hanoi certainly welcomed the American drawdown, several factors contributed to its decision to eschew a diplomatic solution in favor of a military one. First, Hanoi interpreted South Vietnam's deeply flawed Lam Song 719 incursion into Laos in late winter 1971 as proof that Nixon's Vietnamization policy had failed. With the "puppet" army faltering and the Americans on their way out, the time seemed right to strike. Another Nixon policy, however, was deeply alarming to Hanoi. Pacification under the CORDS program had vastly increased Saigon's areas of control in the rural backcountry. The longer the program continued, the politburo worried, the more difficult it would be to reverse. Finally, the U.S. president's overtures to their Soviet and Chinese patrons were also concerning. With Nixon's upcoming visits to Peking in February and Moscow in

May, Hanoi began to suspect that its communist allies were more interested in cultivating improved relations with the United States than in socialist solidarity in Vietnam.

Fearing an impending reduction or even total cutoff of aid and diplomatic support, North Vietnamese General Secretary Le Duan believed the time had come to revive his favored strategy—the general offensive-general uprising. To that end, he and the politburo authorized the Nguyen Hue Offensive, a massive conventional invasion of the South on three fronts. Launched on 30 March, the invasion marshaled the largest and most powerful North Vietnamese force ever assembled, with more than 225,000 troops, hundreds of tanks and armored vehicles, heavy artillery, and anti-aircraft assets poised to strike. The invasion was a stunning success in its opening weeks, as NVA divisions rolled over their outmatched southern counterparts along the DMZ, the Central Highlands, and northwest of Saigon. Only the frenzied close air support by remaining U.S. fighter-bombers and their RVNAF allies temporarily stemmed the tide. But if South Vietnam were to survive, drastic action would be needed.

Furious at Hanoi's provocation, Nixon resolved to make the politically fraught decision to surge massive combat air assets back into Southeast Asia. For the first time in the war, the U.S. would mount an all-out air offensive against targets throughout North Vietnam, including its industrial heartland in the Hanoi-Haiphong complex. Further, Nixon resolved to finally take the advice top military commanders had been issuing for years—mine Haiphong Harbor and other North Vietnamese ports. If Hanoi had settled on a military solution in the South, Nixon decided, then the U.S. would do its level best to devastate the North's warmaking capacity by crippling its infrastructure and strangling the flow of imported supplies. As Operation Linebacker I staunched the flow of men and materiel south, regrouped RVN forces began to halt and then partially roll back Northern gains.

By late summer, Hanoi was forced to concede that its Nguyen Hue campaign had proved a costly failure. While invading forces had acquired about 10 percent of South Vietnam's backcountry, there would be no capitulation by the RVN military and no general uprising among its people. Moreover, the ferocious American response had not only devastated its military, infrastructure, and economy, but threatened the very survival of Hanoi's revolution. Thus, the time had come to reprioritize the diplomatic solution. The objective now was to wrangle an agreement that got the Americans out—and kept them out. Once the North had recovered militarily and economically from the U.S. bombing campaign, eventual victory over the Southern "puppet" was all but assured.

When peace talks resumed in late July, Hanoi instructed its envoys in Paris to drop the longstanding demand that Thieu resign as a part of any final agreement. Their American counterparts were delighted and hastened to offer their own concessions. Over the next two months progress was slow but steady. By October, Kissinger—breaking the administration's repeated promise to fully consult and gain Thieu's approval prior to concluding any settlement— nevertheless finalized a deal with the North Vietnamese. The envoy was convinced he had achieved Nixon's "honorable peace," ending American involvement in Vietnam, preserving U.S. credibility, and providing South Vietnam the means to survive. To help seal the deal, Nixon halted bombing above the 20th parallel and agreed to the signing schedule his envoy had negotiated. All that remained was to gain Thieu's acquiescence.

But the South Vietnamese president, already deeply skeptical of any peace agreement with his hated enemies in the North, was only further enraged when he saw its provisions. The settlement, Thieu believed, was nothing short of surrender. Among its many flaws, it failed to recognize the DMZ as a legitimate political boundary, created a post-ceasefire entity that smacked of coalition government, and placed the Viet Cong's recently created PRG on an equal footing with the RVN. Above all, he most vehemently condemned the agreement for allowing more than 140,000 NVA troops to remain in South Vietnamese territory. All, according to Thieu, delegitimized the very notion of the RVN as an independent, sovereign state—undermining all for which the United States and South Vietnam had sacrificed. No, said, Thieu, he would not sign.

Nixon, facing his own constraints from a hostile Congress and war-weary public, threatened Thieu with a total cutoff of U.S. support if he did not go along. But the South Vietnamese president was unmoved, refusing to be the "puppet" his Northern enemies had long accused him of being. While Nixon and Kissinger were at various times open to signing a bilateral agreement with Hanoi and simply walking away, the demands of an "honorable peace"—and the implications for U.S. credibility in Southeast Asia and beyond— required gaining Thieu's acceptance. So, Kissinger returned to Paris and attempted to walk back several concessions to which he had already agreed. Predicably incensed, Tho accused the American of trying to renegotiate a completed agreement. Over ensuing weeks, the talks became ever-more contentious. Hanoi, increasingly convinced that U.S. domestic and congressional pressure would soon force Nixon to concede, instructed its envoys to dig in their heels. Time, they concluded, was on their side. By mid- December, the talks had broken down completely.

Nixon again faced a politically fraught decision: sign a separate deal and abandon South Vietnam, allow negotiations to fruitlessly drag on—or send the bombers back in. On this last, circumstances had changed dramatically since May. Now, there was no massive North Vietnamese invasion to repel. Months of progress in the talks—punctuated by Kissinger's "peace is at hand" pronouncement in late October—had convinced Congress and the American public that the war was all but over. To resume bombing now risked not just a new round of domestic and international outrage but a complete cutoff of congressional funding for the war. And yet Nixon, now backed into a corner by Hanoi, decided to go for broke and unleash an unprecedented attack on North Vietnam's industrial heartland.

Codenamed Linebacker II, the new campaign included the first mass use of B-52s over North Vietnam. Begun on 18 December and derisively dubbed the "Christmas bombings" by its critics, the campaign saw the big Stratofortresses pummel North Vietnamese military and industrial targets by night, while swarms of Air Force and Navy tactical fighter-bombers unleashed havoc by day. Although the North Vietnamese managed to down 15 of the big bombers over the 11-day campaign, the White House showed no sign of letting up. Finally, with its industrial base wrecked and air defenses all but exhausted, Hanoi was again ready to talk.

But the politburo was not alone in needing a settlement. With public opinion once more inflamed and Congress threatening open revolt, Hanoi's resignation came not a moment too soon for Nixon, as well. Both sides desperate for a deal, the talks proceeded at a blistering pace. Less than a week after resuming negotiations on 8 January, the Americans and North Vietnamese had their settlement. With the end in sight, Nixon resolved to move forward with or without Thieu. He presented the South Vietnamese president with a final ultimatum: trust and consent to what his government had negotiated, and South Vietnam would have his continued support. Reject this final offer, and the RVN was on its own. Out of options and desperate to retain American assistance, Thieu finally relented on 21 January. The fate of his country now rested on whether the United States would honor its commitments. On 27 January, delegates in Paris signed the Agreement on Ending the War and Restoring the Peace in Viet-Nam. Over the next two months, the U.S. pulled the last of its military forces out, while hundreds of American POWs came home to cheering crowds. America's war in Vietnam was over.

Predictably, the settlement ended neither the war nor restored peace in Vietnam. Just prior to the ceasefire, NVA forces in the South launched the three-phase operation conceived the previous fall. The plan called for the rapid capture of as many villages and hamlets possible before the agreement went into effect, thereby expanding areas of communist control in the eyes of the supervisory ICCS. But the South Vietnamese were ready. The resulting "War of the Flags" saw RVN forces repel NVA attacks and recapture many of the villages and hamlets seized in recent days. Bolstered by the influx of U.S. arms and equipment via the Enhance Plus program, South Vietnamese forces were more than holding their own. For the moment, the "balance of forces" between North and South were roughly equal. It would not hold.

Alarmed by Saigon's gains, the politburo in March authorized yet another full-scale invasion. The plan included dramatically upgrading the Ho Chi Minh Trail, complete with all-weather roads and a 3,000-mile pipeline to fuel the coming mechanized assault. By October, U.S. intelligence estimated that North Vietnamese forces in the South had grown by some 70,000 troops, 400 tanks, 200 artillery pieces, and 15 antiaircraft units. Twelve airfields had also been constructed in South Vietnamese territory. All of this, of course, was a flagrant violation of the settlement's prohibition on infiltration and reinforcement. Nixon had repeatedly pledged to "respond massively" to such violations. But events had overtaken his promises. With the war over, POWs home, and U.S. forces withdrawn, the president had neither congressional nor public support for a return to Vietnam. Even if he had, he was not eager to create more prisoners for the Hanoi Hilton.

Earlier that summer, Congress enacted two bills that ensured an end to U.S. military action in Indochina. Nixon's longtime fears of a congressional cutoff of war funding were finally realized on 1 July as legislators passed an appropriations bill forbidding use of funds for military action in Indochina. With the Khmer Rouge communist insurgency pushing hard in Cambodia, Nixon managed to secure congressional approval for continued bombing until 15 August. But when the last B-52 operations in Cambodia ceased on that date, all U.S. military action in Indochina came to an end. Meanwhile, the House and Senate passed the War Powers Resolution by overwhelming majorities that year. The act required the president to notify Congress within 48 hours of U.S. troops engaging in foreign combat, and placed a 60-day limit on any operation without further congressional authorization. Calling the resolution "unconstitutional," Nixon vetoed the legislation on 24 October, only to see his veto overridden by Congress just two weeks later. There would be no "massive reaction" to any North Vietnamese violations. Meanwhile, Kissinger and Tho were jointly awarded the Nobel Peace Prize on 16 October 1973. Tho refused, citing the ongoing war in Vietnam. To avoid antiwar protestors, Kissinger accepted his in absentia, only to try to return it later. The Nobel Committee refused. The Kissinger-Tho award remains one of the most controversial in the Prize's 123-year history. On the agreement's one-year anniversary, nearly 14,000 South Vietnamese soldiers and another 2,100 civilians had been killed following the 27 January "peace" settlement. North Vietnam had lost another 45,000 dead.

With Nixon's influence buckling under the weight of the ever-worsening Watergate scandal, even continued U.S. foreign aid to South Vietnam seemed in jeopardy. On 1 July, Congress slashed funding from the $1.1 billion appropriated for FY 1974 (July 1973- June 1974) to just $750 million for FY 1975. At levels that low, South Vietnam would be unable to stop a major offensive, according to Gen. John Murray, commander of the Defense Attache Office in Saigon. Any hope of reversing the trend evaporated with Nixon's resignation on 9 August 1974. Soon, the reduction in funding necessitated a drastic cutback in fuel, ammunition, equipment maintenance, and more for the South Vietnamese military. Murray repeated his warning that fall, predicting that South Vietnam would capitulate the following year if it did not receive "proper support." By the end of 1974, ARVN soldiers were allotted just 85 rounds of ammunition per month. Meanwhile, a dearth of spare parts meant that an ever-growing portion of South Vietnam's air force and mechanized units would be unavailable for service. At the same time, the Soviet Union, hoping to gain influence in Indochina at the expense of its Chinese rivals, continued aiding North Vietnam to the tune of $1 billion annually. It would help Hanoi assemble yet another massive invasion force. This time, however, there would be neither Nixon nor U.S. airpower to stop it.

In early 1975, Hanoi launched its Campaign 275, a limited operation to seize the Central Highlands and the northern portion of South Vietnam. But NVA commanders soon found themselves in control of vast swaths of new territory, as ARVN resistance melted away under the onslaught. Stunned by the relative ease of the NVA's advance, Hanoi moved up its bid to fully conquer the South, which had originally been planned for 1976. In March, the politburo renamed the offensive the Ho Chi Minh Campaign and pushed forward at full speed. Soon, all of South Vietnam was teetering on the brink.

U.S. President Gerald Ford dispatched Army Chief of Staff Gen. Fredrick Weyand to assess the situation. The general reported that the country might survive if it received an emergency infusion of some $722 million in aid. Ford appealed to congressional leaders, but his pleas fell on deaf ears. Congress and the American people were no longer interested in Vietnam. Instead, Ford ordered Operation Frequent Wind. Over two days, the effort evacuated the remaining Americans and 7,000 South Vietnamese. Ultimately, various U.S. operations would rescue some 130,000 South Vietnamese refugees in 1975. Just before 0800 on 30 April, with communist forces pushing into the capital and desperate throngs of South Vietnamese clamoring at the gates, the last American helicopter lifted away from the embassy roof in Saigon.

The end near, and with several coup plots brewing against him, Thieu had resigned five days earlier, transferring authority to Vice-President Tran Van Huong before fleeing the country. He would eventually make his way to America, where he lived until his death in 2001. Thieu spent the rest of his life deeply bitter over the United States' "betrayal" of his country. As for Huong, he lasted just three day before handing the reins to Gen. Duong "Big Minh" Van Minh. Ironically, Minh was one of the generals who had helped overthrow President Ngo Dinh Diem in November 1963. He was not in charge for long. On 30 April, NVA forces overran Saigon, their Soviet-built T-55 tanks crashing the gates of the Presidential Palace. After more than 20 years of bitter war, Hanoi had finally realized its dream of a unified Vietnam under communist rule.

Bibliography

Books, Interviews, Periodicals, Documents

"2nd Biggest Bombing of North This Year is Reported by U.S." *The New York Times,* October 16, 1972, p. 77.

Agreement on Ending the War and Restoring Peace in Vietnam (The Vietnam Agreement and Protocols, Signed January 27, 1973). Accessed from the Indochina Archive, University of California.

Alvarez, Everett. Interviewed by author, 12/14/21, Transcript 1.

Alvarez, Everett & Pitch, Anthony. *Chained Eagle: The Heroic Story of the First American Shot Down over North Vietnam.* Dulles, VA: Potomac Books, 2005.

Air Force Historical Research Agency, *Air Force After Action Report,* Maxwell Air Force Base, April 1973.

Air Force Historical Research Agency, *Operation Homecoming, Inception Through Implementation, 1964 – 1973,* Maxwell, Air Force Base, 1984.

Andrade, Dale. *America's Last Vietnam Battle: Halting Hanoi's 1972 Easter Offensive.* Lawrence, KS: University Press of Kansas, 2001

Andrade, Dale. *Trial by Fire: The 1972 Offensive, America's Last Vietnam Battle.* New York, NY: Hippocrene Books, 1995

Asselin, Pierre. *A Bitter Peace: Washington, Hanoi, and the Making of the Paris Peace Agreement.* Chapel Hill, NC: The University of North Carolina Press, 2002.

Big Power Plant Raided in North. *The New York Times,* June 12, 1972, p. 1

Boniface, Roger. *MiGs over North Vietnam: The Vietnam People's Air Force in Combat, 1965-75.* Mechanicsburg, PA: Stackpole Books, 2010.

Bowman, John. *The World Almanac of the Vietnam War.* New York, N.Y.: Bison Books, 1985.

Boyne, Walter. *Linebacker II.* Air Force Magazine, November 1997

Brand, Matthew C. *Airpower and the 1972 Easter Offensive.* Master's Thesis. U.S. Army Command and General Staff College. Fort Levenworth, KS. 2007

Browne, Malcolm. "Copters ferry More Saigon troops Into Anloc as enemy fire ebbs." *The New York Times*, June 15, 1972, p. 2.

Browne, Malcolm. "U.S. planes raid 4 bases in North and claim 5 MiG's." *The New York Times*, Oct. 1, 1972, p. 1.

Butterfield, Fox. "Saigon's troops advance within two miles of Anloc." *The New York*

Times, May 18, 1972, p. 1.

Central Intelligence Agency. *North Vietnam: The Dike Bombing Issue*. Intelligence Memorandum. July 1972.

Central Intelligence Agency. *Possible Alternatives to the Rolling Thunder Program*. Intelligence Memorandum. July 1968.

Certain, Robert. Interviewed by author, 12/9/21, Transcript 1; 12/14/21, Transcript 2.

Certain, Robert. *Unchained Eagle: From Prisoner of War to Prisoner of Christ*.

Roswell, GA: Unchained Publications, Kindle Format, 2003.

Chapter XI, Special Forces Association. *Operations Ivory Coast/Kingpin*, 2020.

Correll, John. *The Air Force in the Vietnam War*. The Air Force Association, December 2004.

Correll, John. *Take it Down! The Wild Weasels in Vietnam,* Air & Space Forces Magazine, July 2010.

Cross, Coy. *Operation Homecoming: MAC's Finest Hour*, Beale AFB, CA: 9th Reconnaissance Wing, History Office, 1999.

Davis, Vernon. *The Long Road Home: U.S. Prisoner of War Policy and Planning in Southeast Asia*. Washington, D.C.: Historical Office, Office of the Secretary of Defense, 2000.

DeBellevue, Charles. Interviewed by author, 1/6/22, Transcript 1.

"Defector Tells of Massacre by Enemy at Quang Tri." *The New York Times*.

September 9, 1972, p. 6

Department of the Air Force. *PW Nutrition Care Plan*, Washington, D.C.: 8 Dec 1972.

Drummer, Janene and Wilcoxson, Kathryn. *Chronological History of the C-9A Nightingale*. Scott Air Force Base, IL: Office of History, Air Mobility Command, 2001.

Eiler, Edward. Interviewed by author, 12/28/21, Transcript 1.

Emerson, Stephen. *North Vietnam's 1972 Easter Offensive: Hanoi's Gamble*. South Yorkshire, ENGLAND: Pen and Sword Military, 2020.

Emerson, Stephen. *Vietnam's Final Air Campaign: Operation Linebacker I & II, May-December 1972*. South Yorkshire, England: Pen and Sword Military, 2013.

Ensch, John. Interviewed by author, 1/26/22, Transcript 1.

Eschmann, Karl. *Linebacker: The Untold Story of the Air Raids over North Vietnam*. New York, NY: Ivy Books, 1989

Ethell, Jeffrey & Price, Alfred. *One Day in a Long War: The Greatest Battle of the Vietnam Air War.* London: Silvertail Books, 2019

Evert, Richard. Interviewed by author, 1/17/22, Transcript 1.

Every, Martin and Parker, James. *Aircraft Escape and Survival Experiences of Navy Prisoners of War.* Office of Naval Research, August 1974.

Faram, Mark. "Race riot at sea—1972 *Kitty Hawk* incident fueled fleetwide unrest," *Navy Times.* February 28, 2017.

Finch, Joseph (ed.). *Faces of the Distinguished Flying Cross of Central Florida: In Their Own Words.* The Villages, FL: Finch Publications, 2017.

Francis, Richard. Interviewed by author, 12/17/21, Transcript 1; 12/20/21, Transcript 2.

Frisbee, John L. *The Air War.* Air Force Magazine. September 1972

Fulghum, David and Terrence Maitland. *The Vietnam Experience: South Vietnam on Trial, Mid-1970 to 1972.* Boston: Boston Publishing Company, 1984.

Futrell, R. Frank et al. *The United States Air Force in Southeast Asia, Aces and Aerial Victories, 1965-1973.* The Albert F. Simpson Historical Research Center Air University and Office of Air Force History Headquarters USAF, 1976.

Gonzalez, Michael. *The Forgotten History: The Mining Campaigns of Vietnam, 1967-1973.* War Stories Collection, Angelo State University, TX.

Gwertzman, Bernard. "France's mission in Hanoi wrecked during a U.S. raid." *The New York Times*. October 12, 1972, p. 1.

"Hanoi area raids reported curbed." *The New York Times*. October 13, 1972, p. 11.

Haig, Alexander. *Inner Circles: How American Changed the World: A Memoir*. New York, NY: Grand Central Publishing, 1992.

Hartsook, Elizabeth and Stuart Slade. *Air War: Vietnam Plans and Operations, 1969-1975.* Newton, CT: Defense Lion Publications, 2012.

Hobson, Chris. *Vietnam Air Losses, United States Air Force, Navy and Marine Corps Fixed-Wing Aircraft Losses in Southeast Asia 1961-1973*, 2001, Midland Publishing, Great Britain.

Honour, Craig. Interviewed by author, 1/26/22, Transcript 1.

Hubbell, John. *P.O.W.: A Definitive History of the American Prisoner-of-War Experience in Vietnam, 1964-1973.* New York, NY: Reader's Digest Press, 1976.

Hughes, Ken. *Fatal Politics: The Nixon Tapes, the Vietnam War, and the Casualties of Reelection.* Charlottesville, VA: University of Virginia Press, 2015

Isaacs, Arnold. *Without Honor: Defeat in Vietnam and Cambodia.* Baltimore, MD:

Johns Hopkins University Press, 1983

Isaacson, Walter. *Kissinger: A Biography.* New York: Simon & Schuster, 1992.

Israel, Kenneth. *Operation Homecoming.* United States Air Force, Installation Chaplain. 6 April 1973.

Johnson, Calvin. *Linebacker Operations: September – December 1972.* San Francisco, CA: Project CHECO, Office of History, HQ PACAF, 1973.

Joiner, Gary and Dean, Ashley. *Operation LINEBACKER II: A Retrospective: PART 6: LINEBACKER II.* Report of the LSU Shreveport unit for the SAC Symposium, December 2, 2017

Karnow, Stanley. *Vietnam: A History.* New York: Viking Press, 1983.

"Key rail line reported cut". *The New York Times.* June 15, 1972, p. 2.

Kissinger, Henry. *White House Years.* Boston, MA: Little, Brown and Company, 1979

Kohloff, Arthur. *The BUFs Go Downtown.* "Vietnam" magazine, p. 46-53, 1990

Kowal, Patricia Campbell. *Reflection on Operation Homecoming*, in Aviation Space and Environmental Medicine, 61: 1156-9, 1990.

Lewis, Flora. "Vietnam Peace Pacts Signed; America's Longest War Halts." *The New York Times*, January 28, 1973, p. 1.

Logan, William. *Hanoi: Biography of a City. University of New South Wales* Press, 2000.

Mann, David. *The 1972 Invasion of Military Region I: Fall of Quang Tri and Defense of Hue.* Directorate of Operations Analysis, Project CHECO, 15 March 1973.

Mantel, Mikelene, Interviewed by Author, 1/28/22, Transcript 1. Marolda, Edward. *Ready Seapower: A History of the U.S. Seventh Fleet.*

Washington Navy Yard, DC: Naval History & Heritage Command, Department of the Navy, 2012.

Massimini, Sebastian. Interviewed by Author, 2/15/22, Transcript 1.

McCarthy, Donald. *MiG Killers: A Chronology of U.S. Air Victories in Vietnam 1965-1973.* North Branch, MN: Specialty Press, 2009

McCarthy, James and Allison, George. *Linebacker II: A View from the Rock.* Barksdale Air Force Base, LA: History & Museums Program, Air Force Global Strike Command, 1976/2018.

Michel, Marshall. *Clashes: Air Combat over North Vietnam, 1965-1972.*

Annapolis, MD: Naval Institute Press, 1997.

Middleton, Drew. *Air War – Vietnam.* New York: Arno Press, 1978

United States Military Assistance Command, Vietnam. *1972-1973 Command History Volume I.* Office of the Secretary, Joint Staff Military Assistance Command, Vietnam. 1973

Major field hit. *The New York Times*, May 17, 1972. p. 1.

Melson, Charles & Arnold, Curtis. *U.S. Marines in Vietnam: The War that Would Not End, 1971-1973.* Washington, D.C.: History and Museums Division, Headquarters, United States Marine Corps, 1991

Meredith, Elizabeth. *Jane Fonda: Repercussions of Her 1972 Visit to North Vietnam.* Maxwell Air Force Base, AL: National Geo-Spatial Intelligence Agency, 2010.

Moe, Thomas. Interviewed by Author, 2/9/22, Transcript 1.

Morrocco, John. *The Vietnam Experience: Rain of Fire, Air War 1969-1973.*

Boston: Boston Publishing Company, 1985.

Moyar, Mark. *Triumph Regained: The Vietnam War, 1965-1968.* New York: Encounter Books, 2022.

Nalty, Bernard. *Air War over South Vietnam*, Washington, D.C.: Center of Air Force History, 1995

National Intelligence Council. *Vietnamese Intentions, Capabilities, and Performance Concerning the POW/MIA Issue (U).* Washington, D.C.: Approved for Publication by the National Foreign Intelligence Board, 1998.

Naval History and Heritage Command. *H-Gram 070: The Easter Offensive— Vietnam 1972*, 27 April 2022.

Naval History and Heritage Command. *H-Gram 074: The Easter Offensive— Vietnam 1972 (2), 6 September 2022.*

Nguyen, Lien-Hang T. *Hanoi's War: An International History of the War for Peace in Vietnam.* Chapel Hill, NC:The University of North Carolina Press, 2012.

Nguyen Quy, Editor, *Politburo Resolution No. 194-NQ/TW, 20 November 1969, Van Kien Dang Toan Tap, 30, 1969* [Collected Party Documents, Volume 30, 1969] (Hanoi: National Political Publishing House, 2004), p. 303-305. Translated for CWIHP by Merle L. Pribbenow.

Nixon, Richard. *RN: The Memoirs of Richard Nixon.* New York: Warner Books, 1978.

"North Vietnamese are now requiring everyone to work." *The New York Times*, July 18, 1972, p. 3

Ohlinger, John and Baerenz, Fred. *History of Operation Homecoming, 6 September*

1971 to 27 July 1973. Clark Air Base, R.P.: Headquarters, 13th Air Force.

Pettyjohn, Frank. *The Return of Vietnam POWs—The Aeromedical Phase of Operation Homecoming 1973,* in Aviation Space and Environmental Medicine, 69: 1207-10, 1998.

Phillippe, Jerry. Interviewed by Author, 1/4/22, Transcript 1.

Philpott, Tom. *Glory Denied: The Saga of Vietnam Veteran Jim Thompson, America's Longest-Held Prisoner of War.* New York, NY: Plume, 2002.

Porter, Gareth. *A Peace Denied: The United States, Vietnam, and the Paris Agreement.* Bloomington, IN: Indiana University Press, 1975.

Porter, M. F. *Linebacker: Overview of the First 120 Days.* San Francisco, CA: Project CHECO, Office of History, HQ PACAF, September 1973.

Powers, John. *Treatment of American Prisoners of War in Southeast Asia 1961- 1973.* 2018.

Pribbenow, Merle (Translator). *Victory in Vietnam: The Official History of the People's Army of Vietnam, 1954-1975.* Lawrence, KS: University Press of Kansas, 2002.

Raids near Hanoi are said to wreck big steel works." *The New York Times.* June 26, 1972, p. 1.

Rayman, Russell. *Operation Homecoming: 25 Years Later,* in Aviation Space and Environmental Medicine, 96: 1204-6, 1998.

Reeves, Richard. *President Nixon: Alone in the Whitehouse.* New York, NY: Simon & Schuster, 2001

Rochester, Stuart. *The Battle Behind Bars: Navy and Marine POWs in the Vietnam War.* Washington D.C.: Naval History & Heritage Command, 2010.

Rosenbaum, David. "Efficacy of the Bombing of North Vietnam," *The New York Times.* December 26, 1972, p. 10

Schmitz, David. *Richard Nixon and the Vietnam War: The End of the American Century.* Lanham, MD: Rowman & Littlefield, 2014.

Sheehan, Neil. *A Bright Shining Lie: John Paul Vann and America in Vietnam.* New York, N.Y.: Random House, 1988.

Sherwood, John. *Nixon's Trident: Naval Power in Southeast Asia, 1968-1972.* Washington D.C.: Naval History & Heritage Command, Department of the Navy, 2009.

Sherwood, John. *Fast Movers: Jet Pilots and the Vietnam Experience.* New York, N.Y.: St. Martin's Press, 1999.

Smith, John. *The Linebacker Raids: The Bombing of North Vietnam, 1966-1973.*

London: Arms and Armour Press, 1998.

Solis, Gary. *Marines and Military Law in Vietnam: Trial by Fire.* Washington, D.C.: History and Museums Division Headquarters, U.S. Marine Corps, 1989.

Sorely, Lewis. *A Better War: The Unexamined Victories and Final Tragedy of America's Last Years in Vietnam.* New York: Harcourt Brace & Co., 1999.

Text of Intelligence Report on Bombing of Dikes in North Vietnam Issued by State Department. *The New York Times*, July 29, 1972, p. 2.

Tilford, Earl. *Setup: What the Air Force Did in Vietnam and Why.* Maxwell Air Force Base, AL: Air University Press, 1991.

Thompson, Wayne. *To Hanoi and Back: The U.S. Air Force and North Vietnam, 1966-1973.* Washington, D.C.: Smithsonian Institute Press, 2000.

Thompson, W. Scott and Donaldson D. Frizzell, (ed.), *The Lessons of Vietnam.*

New York, NY, Crane, Russak & Co., 1977

Tiedemann, Mark. Interviewed by Author, 12/7/21, Transcript 1.

Towery, Tommy (ed.). *We Were Crew Dogs I: The B-52 Collection.* Memphis, TN: Tommy Towery, 2019.

Tran Van Don. *Our Endless War: Inside Vietnam.* San Rafael, CA: Presidio Press, 1978.

Treaster, Joseph. "B-52's hit North Vietnam for first time in 7 weeks," *The New York Times*, June 9, 1972, p. 1.

Truong, Ngo Quang. *The Easter Offensive of 1972.* Washington, D.C.: U.S. Army Center of Military History, Indochina Monographs, 1980.

Tucker, Gerald. Interviewed by author, 1/24/22, Transcript 1.

"U.S. copter fire reported to kill South Vietnamese." *The New York Times*, Oct.2, 1972, p. 3.

"U.S. reports bombing of a power plant, blacking out part of Hanoi." *The New York Times*, June 27, 1972, p. 15.

USS *Hancock* (CVA-19) Command History for 1972, FPO San Francisco, August 8, 1973.

USS *Midway* (CVA-41) Command History Calendar Year 1972 Vetter, Dean. Interviewed by author, 12/6/21, Transcript 1.

Welles, Benjamin. "Soviet unit of 8 warships is reported off Vietnam." *The New York Times*, May 17, 1972, p. 19.

Whitfield, Danny. *Historical and Cultural Dictionary of Vietnam*. Metuchen, NJ: The Scarecrow Press, 1976.

Whitney, Craig. "U.S. planes strike at power plants in the Hanoi area." *The New York Times*, 25 May 1972, p. 1.

Willbanks, James. *Thiet Giap! The Battle of An Loc, April 1972*. Combat Studies Institute, U.S. Army Command and General Staff College: Fort Leavenworth, KS, 2011.

Willbanks, James. *Vietnam War Almanac: An In-Depth Guide to the Most Controversial Conflict in American History*, New York, N.Y.: Skyhorse Publishing, 2013.

Williams, Steve. Interviewed by author, 12/15/21, Transcript 1.

Wyatt, Barbara (ed.). *We Came Home.* Toluca Lake, CA: P.O.W. Publications, 1977.

Online/Digital Sources

Address to the Nation Making Public a Plan for Peace in Vietnam – January 25, 1972.

https://www.nixonfoundation.org/2017/09/address-nation-making-public-plan- peace-vietnam-january-25 1972/

AMCOM, *TOW.* U.S. Army Aviation and Missile Life Cycle Management Command. Retrieved August 30, 2024. https://history.redstone.army.mil/miss-tow.html

Bernier, Robert. "The MiG Hunters." *Air & Space Quarterly*. March 22, 2023.

https://airandspace.si.edu/air-and-space-quarterly/spring-2023/mig-hunters#:~:text=On%20May%2023%2C%201972%2C%20McKeown,the%20wildly%20maneuvering%20enemy%20fighters.

Beschloss, Michael. *LBJ and the Descent Into War*. February 5, 2019. https://www.historynet.com/lbj-and-the-descent-into-war/

Branum, Don. *Records Detail MiG Kill by 'Diamond Lil' Tail Gunner.* December 24, 2010.

https://www.af.mil/News/Features/Display/Article/142829/records-detail-mig-kill-by-diamond-lil-tail-gunner/

Busboom, Stanley. *Bat 21: A Case Study*, U.S. Army War College, Carlisle Barracks, PA, April 2, 1990.
https://web.archive.org/web/20120320123131/http://www.dtic.mil/cgi-bin/GetTRDoc?Location=U2&doc=GetTRDoc.pdf&AD=ADA220660

Cable, Dave. *Remembering Kelly Patterson on "Black Friday."* January 28, 2020. https://www.intruderassociation.org/pdf/Kelly-Shoot-Down-1967.pdf

Cagle, Malcolm. *Task Force 77 in Action off Vietnam.* May 1972. https://www.usni.org/magazines/proceedings/1972/may/task-force-77-action- vietnam

Central Intelligence Agency. *The Pros and Cons of Bombing North Vietnam.* June 4 2002

https://www.cia.gov/readingroom/document/cia-rdp78s02149r000100130005-1

Complete List of Returned Prisoners of War, 2201905009. Vietnam Center and Sam Johnson Vietnam Archive. October 1973, Box 19, Folder 05, Douglas Pike Collection: Unit 03 - POW/MIA Issues, Vietnam Center and Sam Johnson Vietnam Archive, Texas Tech University https://www.vietnam.ttu.edu/virtualarchive/items.php?item=2201905009

Cooley, Jason. *Understanding the Failure of the US Security Transfer during the Vietnam War.* Journal of Indo-Pacific Affairs, September 21, 2023 https://www.airuniversity.af.edu/JIPA/Display/Article/3533521/understanding-the-failure-of-the-us-security-transfer-during-the-vietnam-war/

Correll, John. *Take it Down! The Wild Weasels in Vietnam.* July 1, 2010. https://www.airandspaceforces.com/article/0710weasels/

Correll, John. T. *The Emergence of Smart Bombs.* March 1, 2010. https://www.airandspaceforces.com/article/0310bombs/

D'Costa, Ian. *The F-8 Crusader Once Scared a Vietnamese MiG Pilot Into Ejecting Before a Dogfight.*

https://tacairnet.com/2015/04/22/the-f-8-crusader-once-scared-a-vietnamese-mig-pilot-into-ejecting-before-a-dogfight/

D'Costa, Ian. *How Combat Tree Made the F-4 Phantom II the Deadliest Fighter Over Vietnam in the 1970s.*

https://tacairnet.com/2017/01/02/how-combat-tree-made-the-f-4-phantom-ii-the-deadliest-fighter-over-vietnam-in-the-1970s/

DePastino, Todd. *Nixon Announces 'Peace with Honor' and the End of the Vietnam War.* (transcript of Nixon's 23 January speech to the nation) https://veteransbreakfastclub.org/nixon-announces-peace-with-honor-and-the-end- of-the-vietnam-war/

Early and Pioneer Naval Aviators Association. *Ronald E. McKeown, Captain USN (Ret.)*

https://www.epnaao.com/BIOS_files/REGULARS/McKeown-

%20Ronald%20E.pdf

Fratus, Matt. *'Hold and Die'—The Marine Who Became a Legend on Easter Sunday.*

https://www.coffeeordie.com/john-ripley

Gamel, Kim. *'The torture stopped': 1969 brought temporary changes to infamous Hanoi Hilton.* Stars and Stripes, Aug. 15, 2019.

https://www.stripes.com/special-reports/vietnam-stories/1969/the-torture-stopped-1969-brought-temporary-changes-to-infamous-hanoi-hilton-1.593300.

Guttman, Jon. *Army and Navy Pilots Joined Together in a Day of Duels Vietnam.* March 15, 2023.

https://www.historynet.com/day-of-duels-operation-linebacker/

Guttman, Jon. *Unpacking the Myths of the F-8 Crusader in Vietnam*

https://www.historynet.com/f-8-crusader-vietnam/

Guttman, Jon. *USN F-4 Phantom II vs VPAF MIG-17/19.* February 26, 2018. https://www.historynet.com/vietnam-book-review-usn-f-4-phantom-ii-vs-vpaf- mig-17-19/

F-4E Manual. *Heatblur F-4E Phantom II: Emergency Systems* https://f4.manuals.heatblur.se/systems/emergency.html#:~:text=Command%20Selector%20Valve%20Handle,front%20seat%20are%20dual%20ejections.

F-4 Phantom II Fighter.

https://www.boeing.com/history/products/f-4-phantom-ii.page

Farley, Richard. *Ross Perot's Forgotten Mission During the Vietnam War.* https://time.com/5759015/ross-perots-forgotten-mission-vietnam-war/

Friedman, Norman. *Armaments and Innovation—The 70 Year History of the Sparrow Missile.* Naval History, Volume 31, Number 6. December 2017 https://www.usni.org/magazines/naval-history- magazine/2017/december/armaments-and-innovation-70-year-history-sparrow#:~:text=In%20addition%2C%20the%20motor%20of,There%20were%20also%20reliability%20problems.

Further reports on Jane Fonda's activities in DRV, 2360309060. Vietnam Center and Sam Johnson Vietnam Archive. 19 July 1972, Box 03, Folder 09, Douglas Pike Collection: Unit 08 - Biography, Vietnam Center and Sam Johnson Vietnam Archive, Texas Tech University https://www.vietnam.ttu.edu/virtualarchive/items.php?item=2360309060

Haley, Heather. *"Shoot 'Em Up": Operation Lion's Den, 27 August 1972.* August 24, 2022

https://usnhistory.navylive.dodlive.mil/Recent/Article- View/Article/3137625/shoot-em-up-operation-lions-den-27-august-1972/

Heck, Timothy. *U.S. Options to Respond to North Vietnam's 1973 Violations of the Paris Peace Accords.* Real Clear Defense, October 21, 2019 https://www.realcleardefense.com/articles/2019/10/21/us_options_to_respond_to_north_vietnams_1973_violations_of_the_paris_peace_accords_114801.html

Hughes, Ken. *Vietnamization.*

https://millercenter.org/the-presidency/educational-resources/vietnamization

Jane Fonda Statement to US Pilots, 2150913020. Vietnam Center and Sam Johnson Vietnam Archive. 24 July 1972, Box 09, Folder 13, Douglas Pike Collection: Unit 03 - Antiwar Activities, Vietnam Center and Sam Johnson Vietnam Archive, Texas Tech University https://www.vietnam.ttu.edu/virtualarchive/items.php?item=2150913020

GlobalSecurity.org. *Operation Freedom Train, Operation Linebacker*

https://www.globalsecurity.org/military/ops/linebacker-1.htm.

GlobalSecurity.org. *KA-3B/EKA-3B*
https://www.globalsecurity.org/military/systems/aircraft/ka-3b.htm#:~:text=Unusually%20large%20for%20a%20carrier,3%2C350%20gallons

%2C%20or%2021%2C775%20pounds

Joint Statement Following Discussions With Leaders of the People's Republic of China. Office of the Historian, Foreign Relations of the United States, 1969-1976, Volume XVII, China, 1969-1972
https://history.state.gov/historicaldocuments/frus1969-76v17/d203

Kort, Michael. *The Vietnam War Reexamined.* Fifteen Eighty Four, Academic Perspectives from Cambridge University Press. November 16, 2017
https://cambridgeblog.org/2017/11/the-vietnam-war-reexamined/

Lam Son 719: Defense POW/MIA Accounting Agency https://dpaa-

mil.sites.crmforce.mil/dpaaFamWebInLamSon#:~:text=U.S.%20forces%20suffered%20a%20staggering,1%2C149%20WIA%2C%20and%2042%20MIA.

Leone, Dario. *Face Curtain vs. Lower Handle.* https://theaviationgeekclub.com/face-curtain-vs-lower-handle- former-us-naval- aviator-explains-why-some-aircraft-have-2-ejection-handles-above-the-seat-while- some-have-a-single-ejection-handle-bet/. 13 December 2022

Leone, Dario. *Here's how B-52 Tail Gunners shot down two North Vietnamese MiG-21 Fighters and turned the iconic strategic bomber into a MiG Killer.*
https://theaviationgeekclub.com/heres-how-b-52-tail-gunners-shot-down-two- north-vietnamese-mig-21-fighters-and-turned-the-iconic-strategic-bomber-into-a- mig-killer/ Dec. 17, 2022

Leone, Dario. *Loose Deuce vs Fluid Four.*

https://theaviationgeekclub.com/loose-deuce-vs-fluid-four-during-the-vietnam- war-the-fighting-tactics-used-by-us-naval-aviators-were-better-than-those-of-the- usaf-pilots-heres-why/

Leone, Dario. *Vietnamese MiG-17 pilot who punched out when he found his opponent was an F-8 instead of an F-4.*

https://theaviationgeekclub.com/the-controversial-story-of-the-north-vietnamese-mig-17-pilot-whopunched-out-when-he-found-his-opponent-was-an-f-8-instead- of-an-f-4/. 7 Nov. 2021

Locher Survival Briefing: Part 1, f4phantom.com https://www.f4phantom.com/docs/oyster1b_pt1.mp3

Locher Survival Briefing: Part 2, Part 1, f4phantom.com https://www.f4phantom.com/docs/oyster1b_pt2.mp3

Marolda, Edward. *Operation Linebacker: The Sea-Power Factor.* U.S. Naval Institute, August 2022.

https://www.usni.org/magazines/naval-history-magazine/2022/august/operation-linebacker-sea-power-factor

Military Times. *Capt. Dale Stovall, Air Force Cross.* https://www.militarytimes.com/citations-medals-awards/recipient.php?recipientid=3468

National Museum of the United States Air Force. *Mikoyan-Guervich MiG-17F.* https://www.nationalmuseum.af.mil/Visit/Museum-Exhibits/Fact-Sheets/Display/Article/196057/mikoyan-gurevich-mig-17f/

National Museum of the United States Air Force. *SA-2 Surface-to-Air Missile.*

https://www.nationalmuseum.af.mil/Visit/Museum-Exhibits/Fact-Sheets/Display/Article/196037/sa-2-surface-to-air-missile/

National Museum of the United States Air Force. *General Dynamics F-111A Aardvark.*

https://www.nationalmuseum.af.mil/Visit/Museum-Exhibits/Fact-Sheets/Display/Article/196049/general-dynamics-f-111a-aardvark/

Naval History and Heritage Command. *The Human Cost.* June 24, 2024. https://www.history.navy.mil/browse-by-topic/wars-conflicts-and-operations/vietnam-war0/human-cost.html#:~:text=A%20few%20figures%20from%20official,per%20service)%20died%20in%20captivity.

Nixon, Richard. *January 25, 1972: Address to the Nation on Plan for Peace in Vietnam.* UVA Miller Center, Presidential Speeches. https://millercenter.org/the-presidency/presidential-speeches/january-25-1972- address-nation-plan-peace-vietnam

Office of the Historian (OOTH): *6. Address by President Nixon, Washington, March 29, 1973* https://history.state.gov/historicaldocuments/frus1969-76v38p1/d6

Office of the Historian: *73. Editorial Note (on Kissinger's 26 October Press Conference)*

https://history.state.gov/historicaldocuments/frus1969-76v09/d73

Office of the Historian: *Breakdown of Negotiations, November 1972–December 1972*

https://history.state.gov/historicaldocuments/frus1969-76v42/ch5

- 26. Memorandum of Conversation: Paris, November 20, 1972, 10:45 a.m.–4:55 p.m.
- 27. Memorandum of Conversation: Paris, November 21, 1972, 3:02–7:26 p.m.
- 32. Memorandum of Conversation: Paris, December 4, 1972, 10:30 a.m.–1 p.m.
- 34. Memorandum of Conversation: Paris, December 6, 1972, 10:40 a.m.–3:50 p.m.
- 35. Memorandum of Conversation: Paris, December 7, 1972, 3–7 p.m.
- 36. Memorandum of Conversation: Paris, December 8, 1972, 3:05–7:20 p.m.
- 38. Memorandum of Conversation: Paris, December 11, 1972, 3:10–7:15 p.m.
- 40. Memorandum of Conversation: Paris, December 12, 1972, 3:07–7:35 p.m.
- 41. Memorandum of Conversation: Paris, December 13, 1972, 10:30 a.m.–4:24 p.m.

Office of the Historian: *Settlement Accomplished: The Accords Initialed and Signed, January 1973*

https://history.state.gov/historicaldocuments/frus1969-76v42/ch6

- 43. Memorandum of Conversation: Paris, January 9, 1973, 9:58 a.m.–3:45 p.m.
- 45. Memorandum of Conversation: Paris, January 11, 1973, 10 a.m.–4 p.m.

Office of the Historian: *111. Memorandum from Secretary of Defense Laird to President Nixon, Washington, November 17, 1972* https://history.state.gov/historicaldocuments/frus1969-76v09/d111

Office of the Historian: 189. *Letter from President Nixon to South Vietnamese*

President Thieu, Washington, December 17, 1972. https://history.state.gov/historicaldocuments/frus1969-76v09/d189

Office of the Historian: *Serious Negotiations and the October Settlement, July 1972-October 1972*

https://history.state.gov/historicaldocuments/frus1969-76v42/ch4

- 21. Memorandum of Conversation: Paris, October 8, 1972, 10:30 a.m. – 7:38 p.m.
- 23. Memorandum of Conversation: Paris, October 10, 1972, 4–9:55 p.m.
- 24. Memorandum of Conversation: Paris, October 11–12, 1972, 9:50 a.m.–2 a.m.
- 25. Memorandum of Conversation: Paris, October 17, 1972, 10:37 a.m.–10:10 p.m.

Office of the Historian: *206. Backchannel Message From the President's Deputy Assistant for National Security Affairs (Haig) to the President's Assistant for National Security Affairs (Kissinger), Saigon, December 20, 1972, 1020Z.* https://history.state.gov/historicaldocuments/frus1969-76v09/d206

Office of the Historian: *254. Backchannel Message From the Ambassador to Vietnam (Bunker) to the President's Assistant for National Security Affairs (Kissinger), Saigon, January 7, 1973, 0505Z.* (letter from Thieu to Nixon)https://history.state.gov/historicaldocuments/frus1969-76v09/d254

Office of the Historian: *256. Message From the President's Assistant for National Security Affairs (Kissinger) to President Nixon, Paris, January 9, 1973, 1620Z.* https://history.state.gov/historicaldocuments/frus1969-76v09/d256

Office of the Historian: *290. Backchannel Message From the President's Assistant for National Security Affairs (Kissinger) to the Ambassador to Vietnam (Bunker), Washington, January 17, 1973, 2345Z* (letter from Nixon to Thieu) https://history.state.gov/historicaldocuments/frus1969-76v09/d290

Office of the Historian: *310. Backchannel Message From the Vice Chief of Staff of the Army (Haig) to the President's Assistant for National Security Affairs (Kissinger), Saigon, January 20, 1973, 0825Z* (letter from Thieu to Nixon) https://history.state.gov/historicaldocuments/frus1969-76v09/d310

Office of the Historian: *313. Backchannel Message From the President's Assistant for National Security Affairs (Kissinger) to the Ambassador to Vietnam (Bunker), Washington, January 20, 1973, 1912Z* (letter from Nixon to Thieu) https://history.state.gov/historicaldocuments/frus1969-76v09/d313

Office of the Historian: *320. Letter From South Vietnamese President Thieu to President Nixon, Saigon, January 21, 1973* https://history.state.gov/historicaldocuments/frus1969-76v09/d320

Office of the Historian: *325. Backchannel Message From the President's Assistant for National Security Affairs (Kissinger) to the Ambassador to Vietnam (Bunker), Washington, January 22, 1973, 0203Z* (letter from Nixon to Thieu) https://history.state.gov/historicaldocuments/frus1969-76v09/d325

Office of the Historian: *341. Telegram From the U.S. Delegation to the Paris Peace Talks to the Department of State, Paris, January 27, 1973, 2308Z* https://history.state.gov/historicaldocuments/frus1969-76v09/d341

Pacific Airlifter. *Prisoner Exchange – Loc Ninh South Vietnam Monday – 12 February 1973.*

https://pacificairlifter.com/prisoner-exchange-loc-ninh-12-feb-1973/

Parks, W. Hays, *Linebacker and the Law of War,* Air University Review, January-February 1983.

https://biotech.law.lsu.edu/cases/nat-sec/Vietnam/Linebacker-and-the-Law-of-War.html

Patterson, Lisa. *Hidden History: The Story of the Indominable Wives and Families Who Fought to Bring POW/MIA Loved Ones Home.* Davidson College, November 10, 2020.

https://www.davidson.edu/news/2020/11/10/indomitable-wives-fought-bring-pow-mia-loved-ones-home

P.O.W. Network: You Are Not Forgotten

- Harris, Jeffrey: https://www.pownetwork.org/bios/h/h149.htm
- Hegdahl, Douglas: https://www.pownetwork.org/bios/h/h135.htm
- Hendrix, Jerry: https://pownetwork.org/bios/h/h179.htm
- Lodge, Robert: https://www.pownetwork.org/bios/l/l068.htm
- Shumaker, Robert: https://www.pownetwork.org/bios/s/s097.htm
- Shott, Richard: https://www.pownetwork.org/bios/s/s198.htm
- Sullivan, Farrell: https://www.pownetwork.org/bios/s/s201.htm
- Thompson, Floyd: https://www.pownetwork.org/bios/t/t030.htm

Presidential Recordings Digital Edition. *Richard Nixon and Henry A. Kissinger on 3 August 1972*. Miller Center, University of Virginia https://prde.upress.virginia.edu/conversations/4006748

Raytheon (Philco/General Electric) AAM-N-7/GAR-8/AIM-9 *Sidewinder*. https://www.designation-systems.net/dusrm/m-9.html

Schuster, Carl: *Arsenal | The SA-7 Grail.*

https://www.historynet.com/sa-7-grail-man-portable-missile-packs-punch/

Schuster, Carl: *North Vietnam's AT-3 Sagger Anti-Tank Missiles.* https://www.historynet.com/north-vietnams- -3-sagger-anti-tank-missiles/

Schuster, Carl: *Terror in the Skies: North Vietnam's Light Anti-Aircraft Artillery.*

https://www.historynet.com/north-vietnams-light-anti-aircraft-artillery/

Society of Wild Weasels: *We Remember. https://wildweasels.org/we-remember/*

Son Bach: *Kham Thien Bombing 50 Years On: Resurrection from Ruins,*

December 2022

https://special.nhandan.vn/kham-thien-bombing-50-years-en/index.html

Tayabji, Jen. *Remembering the Christmas Bombing.* Vietnam Veterans Against the War, December 2012.

https://www.vvaw.org/veteran/article/?id=2204&print=yes

The Vietnam Agreement and Protocols- signed January 27, 1973., 2201905015. Vietnam Center and Sam Johnson Vietnam Archive. 1973, Box 19, Folder 05, Douglas Pike Collection: Unit 03 - POW/MIA Issues, Vietnam Center and Sam Johnson Vietnam Archive, Texas Tech University

https://www.vietnam.ttu.edu/virtualarchive/items.php?item=2201905015

The Wall of Valor Project: The United States of America Vietnam War Commemoration Honoring Service Valor Sacrifice

- *Randall Harold Cunningham.*

https://valor.militarytimes.com/hero/4194

- *William Patrick Driscoll.*

https://valor.militarytimes.com/hero/4195

Tragedy Aboard the USS Stoddert, June 26-August 8.

https://www.vietnamwar50th.com/education/week_of_june_27_2021/

USAF Crash Rescue Team's Fight to Save F-105 Aviators, May 17, 1972

https://www.vietnamwar50th.com/education/week_of_may_16_2021/

The Wall of Faces: Vietnam Veterans Memorial Fund

Farrell Junior Sullivan: https://www.vvmf.org/Wall-of-Faces/50499/FARRELL-J-SULLIVAN/page/2/

Time Magazine. *Viet Nam: The Battle of the Dikes.* August 7, 1972 https://time.com/archive/6816854/viet-nam-the-battle-of-the-dikes/

U.S. Department of Veterans Affairs: *Vietnam War Veterans: Honoring Those Who Served*, 2021

https://www.data.va.gov/stories/s/Vietnam-Veterans-Memorial-Day-2021/q3fu7ckx/#:~:text=The%20Vietnam%20Veterans%20Memorial%20in,also%20wounded%20during%20the%2war.

Veteran Tributes: Honoring Those Who Served

- DeBellevue, Charles: http://veterantributes.org/TributeDetail.php?recordID=128
- Feinstein, Jeffrey: http://veterantributes.org/TributeDetail.php?recordID=1248
- Ferguson, Walter: http://www.veterantributes.org/TributeDetail.php?recordID=506
- Francis, Richard: http://www.veterantributes.org/TributeDetail.php?recordID=1246
- Hegdahl, Douglas: http://www.veterantributes.org/TributeDetail.php?recordID=166
- Lodge, Robert: http://www.veterantributes.org/TributeDetail.php?recordID=148
- Rissi, Donald: http://www.veterantributes.org/TributeDetail.php?recordID=508
- Ritchie, Steve: http://veterantributes.org/TributeDetail.php?recordID=801
- Shumaker, Robert: http://veterantributes.org/TributeDetail.php?recordID=143
- Thomas, Robert: http://veterantributes.org/TributeDetail.php?recordID=507
- Thompson, James Floyd: http://www.veterantributes.org/TributeDetail.php?recordID=300

Vietnam: A Television History; Peace is at Hand (1968 - 1973): The Vietnam Collection; *Interview with John D. Negroponte, 1981* https://openvault.wgbh.org/catalog/V_B828B3CEB0B14ABA823885383D447E99

Vietnam War U.S. Military Fatal Casualty Statistics: https://www.archives.gov/research/military/vietnam- war/casualty-statistics

Villard, Erik. "A Controversial Question: Did Tet Decimate the Vietcong?" March 13, 2021.

https://www.historynet.com/villard-tet/

White, Jeremy Patrick: "Civil Affairs in Vietnam," Center for Strategic and International Studies.http://csis.org/files/media/csis/pubs/090130_vietnam_study.pdf

Notes

Chapter 1

[1] S. Emerson, *1972 Easter Offensive*, p. 7-10; Vietnam War U.S. Fatal Casualty Statistics

[2] Ibid

[3] Nixon Address to the Nation, 25 January 1972; P. Asselin, *A Bitter Peace,* p. 18

[4] Nixon Address to the Nation, 25 January 1972; P. Asselin, *A Bitter Peace,* p. 32- 33; D. Schmitz, *Richard Nixon,* p. 138

[5] P. Asselin, *A Bitter Peace,* p. 34; L. Nguyen, *Hanoi's War,* p. 238

[6] V. Davis, *Long Road Home,* p. 197

[7] V. Davis, *Long Road Home,* p. 197-198

[8] S. Karnow, *Vietnam: A History,* p. 655

[9] V. Davis, *Long Road Home,* p. 206-209

[10] D. Schmitz, *Richard Nixon,* p. 135-138

[11] L. Nguyen, *Hanoi's War,* p. 238-239, 242-243

[12] Ibid

[13] Ibid; M. Moyar, *Triumph Regained,* p.291-292

[14] L. Nguyen, *Hanoi's War,* p. 239-244; E. Villard, *A Controversial Question*

[15] L. Nguyen, *Hanoi's War,* p. 241-244; P. Asselin, *A Bitter Peace,* p. 37-38

Chapter 2

[1] M. Pribbenow (translator), *Victory in Vietnam,* p. 283

[2] Lam Son 719; S. Emerson, *1972 Easter Offensive*, p. 11-12; P. Asselin, *A Bitter Peace,* p. 27

[3] Lam Son 719; P. Asselin, *A Bitter* Peace, p. 27; M. Pribbenow (translator), *Victory in Vietnam,* p. 283-284

[4] M. Pribbenow (translator), *Victory in Vietnam,* p. 283-284

[5] J. White, *Civil Affairs in Vietnam*, p. 10; J. Sherwood, *Nixon's Trident,* p. 35

[6] P. Asselin, *A Bitter Peace,* p. 49-50

[7] M. Pribbenow (translator), *Victory in Vietnam,* p. 284; P. Asselin, *A Bitter Peace,* p. 29, 37-38; S. Emerson, *1972 Easter Offensive*, p. 20

[8] C. Schuster, *North Vietnam's AT-3*

[9] C. Schuster, *The SA-7 Grail*

[10] M. Pribbenow (translator), *Victory in Vietnam,* p. 285

[11] M. Pribbenow (translator), *Victory in Vietnam,* p. 285-286

[12] M. Pribbenow (translator), *Victory in Vietnam,* p. 289

[13] Ibid

[14] Ibid; D. Fulgham and T. Maitland, p. 120-122

Chapter 3

[1] S. Emerson, *1972 Easter Offensive*, p. 21-24

[2] Ibid; M. Fratus, *Hold and Die*; S. Emerson, *1972 Easter Offensive*, p. 46-47; D. Mann, *Fall of Quang Tri,* p. 41

[3] N. Truong, *The Easter Offensive of 1972,* p. 46; The New York Times, *Defector tells of massacre*, p.6

[4] N. Truong, *The Easter Offensive of 1972,* p. 48-50

[5] D. Andrade, *America's Last Vietnam Battle,* p. 145; N. Truong, *The Easter Offensive of 1972,* p. 61

[6] S. Emerson, *1972 Easter Offensive*, p. 26-27

[7] S. Emerson, *1972 Easter Offensive*, p. 27-29; M. Pribbenow (translator), *Victory in Vietnam,* p. 295; Richard Shott, P.O.W. Network Bio

[8] S. Emerson, *1972 Easter Offensive*, p. 30-33; M. Pribbenow (translator), *Victory in Vietnam,* p. 295-296

[9] M. Pribbenow (translator), *Victory in Vietnam,* p. 293-297; S. Emerson, *1972 Easter Offensive*, p. 33-37

[10] P. Asselin, *A Bitter Peace,* p. 44

[11] Ibid

[12] P. Asselin, *A Bitter Peace,* p. 44-45; W. Isaacson, *Kissinger*, p. 417

[13] Air Force Museum, *Mikoyan-Guervich MIG-17F;* Boeing, *F-4 Phantom II Fighter;* J. Guttman, *F-4 vs MIG-17*

[14] J. Ethell & A. Price, *One Day in a Long War,* p. 11-12

[15] E. Hartsook and S. Slade, *Air War,* p. 221; P. Asselin, *A Bitter Peace,* p. 45

[16] R. Nixon, *Memoirs of Richard Nixon,* p. 535, 602

[17] J. Sherwood, *Nixon's Trident,* p. 45

[18] P. Asselin, *A Bitter Peace,* p. 45-48

[19] J. Morrocco, *Air War 1969-1973*, p. 102

[20] M. Beschloss, *LBJ and the Descent Into War;* M. Kort, *Vietnam War Reexamined*

[21] M. Porter, *The First 120 Days,* p. 2-8; S. Karnow, *Vietnam,* p. 591

[22] CIA, *Bombing North Vietnam,* p. 1-4

[23] P. Asselin, *A Bitter Peace,* p. 48; R. Nixon, *Memoirs of Richard Nixon,* p. 607

[24] J. Correll, *Vietnam War Almanac,* p. 56-58; M. Porter, *The First 120 Days,* p. 5;

S. Emerson, *Vietnam's Final Air Campaign,* p. 62; C. Schuster, *Terror in the Skies;* R. Futrell et al, *Aces and Aerial Victories,* p. 83-85; J. Ethell & A. Price, *One Day in a Long War,* p. 20-21

[25] Ibid

[26] Ibid

[27] Ibid

[28] Ibid

[29] J. Morrocco, *Air War 1969-1973*, p. 108-109; E. Hartsook and S. Slade, *Air War,* p. 290-291; D. Middleton, *Air War – Vietnam*, p. 125; S. Emerson, *Vietnam's Final Air Campaign,* p. 89

[30] J. Ethell & A. Price, *One Day in a Long War,* p. 19-20; I. D'Costa, *Combat Tree*

[31] Ibid

[32] J. Ethell & A. Price, *One Day in a Long War,* p. 17-18; J. Correll, *Smart Bombs*;

S. Emerson, *Vietnam's Final Air Campaign,* p. 55-56

[33] J. Gutman, *Unpacking the Myths; Hancock* 1972 Command History, p. 6; R. Evert, Author Interview, Transcript 1

[34] G. Tucker, Author Interview, Transcript 1; I. D'Costa, *F-8 Crusader*; *Hancock*

1972 Command History, p. 6

[35] G. Tucker, Author Interview, Transcript 1

[36] S. Emerson, *1972 Easter Offensive*, p. 40-41

[37] S. Emerson, *1972 Easter Offensive*, p. 43-44

[38] D. Leone, *B-52 Tail Gunners;* M. Tiedemann, Author Interview, Transcript 1

[39] S. Emerson, *1972 Easter Offensive*, p. 42-43

[40] J. Correll, *Smart Bombs;* M. Porter, *The First 120 Days,* p. 24

[41] P. Asselin, *A Bitter Peace,* p. 49-50; Joint Statement; J. Sherwood, *Nixon's Trident,* p. 46

[42] E. Marolda, *Ready Seapower,* p. 62; S. Emerson, *Vietnam's Final Air Campaign,*

p. 44-47; J. Sherwood, *Nixon's Trident,* p. 47-50; MACV, *1972-1973 Command History*, p. B-12; W. Thompson, *To Hanoi and Back,* p. 238-239

[43] S. Emerson, *Vietnam's Final Air Campaign,* p. 44-47; E. Marolda, *Ready Seapower,* p. 64; MACV, *1972-1973 Command History*, p. B-63; M. Moyar, *Triumph Regained, 182;* J. Sherwood, *Nixon's Trident,* p. 47-50

[44] S. Emerson, *Vietnam's Final Air Campaign,* p. 47-48; J. Sherwood, *Nixon's Trident,* p. 50-51; J. Ethell & A. Price, *One Day in a Long War,* p. 15

[45] S. Emerson, *Vietnam's Final Air Campaign,* p. 47-48; J. Sherwood, *Nixon's Trident,* p. 50-51

Chapter 4

[1] S. Emerson, *Vietnam's Final Air Campaign,* p. 78; J. Ethell & A. Price, *One Day in a Long War,* p. 24; E. Marolda, *The Sea-Power Factor*

[2] J. Ethell & A. Price, *One Day in a Long War,* p. 35-39; S. Emerson, *Vietnam's Final Air Campaign,* p. 48-49

[3] Ibid

[4] Ibid

[5] J. Ethell & A. Price, *One Day in a Long War,* p. 40-43, 149; S. Emerson, *Vietnam's Final Air Campaign,* p. 49, 62; J. Sherwood, *Nixon's Trident,* p. 54, 59 [6] J. Ethell & A. Price, *One Day in a Long War,* p. 40-43; S. Emerson, *Vietnam's Final Air Campaign,* p. 49; J. Sherwood, *Nixon's Trident,* p. 54, 59

[7] Ibid

[8] J. Ethell & A. Price, *One Day in a Long War,* p. 50-52; J. Guttman, *Day of Duels*

[9] S. Emerson, *Vietnam's Final Air Campaign,* p. 49; J. Ethell & A. Price, *One Day in a Long War,* p. 56-58; J. Sherwood, *Fast Movers,* p. 219; J. Guttman, *Day of Duels;* C. DeBellevue, Author Interview, Transcript 1

[10] J. Ethell & A. Price, *One Day in a Long War,* p. 56-58; N. Friedman, *Armaments and Innovation;* J. Guttman, *Day of Duels*

[11] R. Futrell et al, *Aces and Aerial Victories,* p. 93; J. Guttman, *Day of Duels;* Dick Francis, Author Interview, Transcript 1; J. Ethell & A. Price, *One Day in a Long War,* p. 59-61

[12] J. Ethell & A. Price, *One Day in a Long War,* p. 62

[13] C. DeBellevue, Author Interview, Transcript 1; J. Ethell & A. Price, *One Day in a Long War,* p. 61

[14] Lodge, Robert. Veteran Tributes; Locher Survival Briefing: Part 1; Locher Survival Briefing: Part 2; J. Guttman, *Day of Duels*

[15] S. Busboom, *Bat 21: A Case Study,* p. 30; P.O.W. Network, *Lodge*; Military Times, *Capt. Dale Stovall*

[16] R. Futrell et al, *Aces and Aerial Victories,* p. 93-104; Veteran Tributes, *Steve Ritchie:* Veteran Tributes, *Charles DeBellevue*; C. DeBellevue, Author Interview, Transcript 1; Veteran Tributes, *Jeffrey Feinstein*

[17] S. Emerson, *Vietnam's Final Air Campaign,* p. 50; J. Ethell & A. Price, *One Day in a Long War,* p. 17

[18] S. Emerson, *Vietnam's Final Air Campaign,* p. 50; J. Ethell & A. Price, *One Day in a Long War,* p. 51, 63-65; MACV, *1972-1973 Command History*, p. B-8

[19] Ibid; Dick Francis, author interview, transcript 1

[20] J. Ethell & A. Price, *One Day in a Long War,* p. 75-76; MACV, *1972-1973 Command History*, p. B-8; S. Emerson, *Vietnam's Final Air Campaign,* p. 50-51 [21] J. Ethell & A. Price, *One Day in a Long War,* p. 82-85, 149; MACV, *1972-1973 Command History*, p. B-8; S. Emerson, *Vietnam's Final Air Campaign,* p. 53, 70;

M. Michel, *Clashes,* p. 240

[22] J. Ethell & A. Price, *One Day in a Long War,* p. 86-88; MACV, *1972-1973 Command History*, p. B-8

[23] J. Guttman, *Day of Duels*; Ethell & A. Price, *One Day in a Long War,* p. 86-88; Raytheon, *Sidewinder*; P.O.W. Network, *Harris, Jeffrey*

[24] S. Emerson, *Vietnam's Final Air Campaign,* p. 53; J. Guttman, *Day of Duels;*

Ethell & A. Price, *One Day in a Long War,* p. 99

[25] J. Sherwood, *Nixon's Trident,* p. 56; S. Emerson, *Vietnam's Final Air Campaign,* p. 53; *One Day in a Long War,* p. 102

[26] J. Sherwood, *Nixon's Trident,* p. 57-58; J. Guttman, *Day of Duels*

[27] Ibid

[28] Ibid

[29] J. Sherwood, *Nixon's Trident,* p. 59; J. Guttman, *Day of Duels*; Ethell & A. Price, *One Day in a Long War,* p. 130

[30] Wall of Valor, *Cunningham*; Wall of Valor, *Driscoll*

[31] J. Guttman, *Day of Duels*

[32] J. Guttman, *Day of Duels*; Ethell & A. Price, *One Day in a Long War,* p. 126- 129; S. Emerson, *Vietnam's Final Air Campaign,* p. 53

[33] Ethell & A. Price, *One Day in a Long War,* p. 148-152; S. Emerson, *Vietnam's Final Air Campaign,* p. 61-62; J. Correll, *Air Force in the Vietnam War,* p. 14

[34] S. Emerson, *Vietnam's Final Air Campaign,* p. 67

[35] Ibid

[36] Ibid; W. Logan, *Hanoi Biography,* p. 67-68

[37] S. Emerson, *Vietnam's Final Air Campaign,* p. 53-54; Ethell & A. Price, *One Day in a Long War,* p. 110, 157 & 163

Chapter 5

[1] S. Emerson, *Vietnam's Final Air Campaign,* p. 51; C. Hobson, *Vietnam Air Losses,* p. 235; J. Morrocco, *Air War 1969-1973*, p. 136

[2] C. Honour, Author Interview, Transcript 3

[3] C. Honour, Author Interview, Transcript 3; J. Bowman, *Almanac of Vietnam War,* p. 311

[4] The New York Times, *Major field hit*, p. 1.

[5] S. Emerson, *Vietnam's Final Air Campaign,* p. 59; *C. Whitney, U.S. planes strike at power plants,* p. 1; R. Futrell et al, *Aces and Aerial Victories,* p. 94

[6] USS *Midway* Command History 1972, p. 7; R. Bernier, *MiG Hunters*

[7] Ibid

[8] J. Ensch, Author Interview, Transcript 1; R. Bernier, *MiG Hunters*

[9] J. Ensch, Author Interview, Transcript 2

[10] Ibid

[11] Ibid

[12] Ibid

[13] Ibid

[14] J. Ensch, Author Interview, Transcript 2; R. Bernier, *MiG Hunters*

[15] J. Ensch, Author Interview, Transcript 2

[16] Early and Pioneer Naval Aviators Association, *Ronald E. McKeown*; J. Ensch, Author Interview, Transcript 1; R. Bernier, *MiG Hunters*

[17] B. Welles, *Soviet unit of 8 warships,* p. 19; M. Moyar, *Triumph Regained,* p. 239

[18] M. Porter, *The First 120 Days,* p. 28; M. Moyar, *Triumph Regained,* p. 240

[19] J. Morrocco, *Air War 1969-1973*, p. 136; The New York Times, *Big Powerplant Raided,* p. 1; E. Hartsook and S. Slade, *Air War,* p. 298; J. Sherwood, *Nixon's Trident,* p. 61; M. Porter, *The First 120 Days,* p. 27; J. Bowman, *Almanac of Vietnam War,* p. 315-116

[20] J. Sherwood, *Nixon's Trident,* p. 64, 66; F. Butterfield, *Saigon's troops advance*, p. 1.

D. Middleton, *Air War – Vietnam*, p. 252

[21] C. Hobson, *Vietnam Air Losses,* p. 47

[22] J. Bowman, *Almanac of Vietnam War,* p. 314

[23] Ibid; J. Treaster, *B-52s hit North Vietnam,* p. 1; F. Butterfield, *Foe attacks town,* p. 1.

[24] *Big Power Plant Raided in North,* p. 1; W. Parks, *Linebacker and the Law*;

[25] M. Browne. *Copters ferry More Saigon troops,* p. 2; The New York Times, *Key rail line reported cut,* p. 2

[26] M. Browne. *Airfields are hit as heavy strikes continue in North*, p. 1.; S. Emerson, *Vietnam's Final Air Campaign,* p. 70-71; The New York Times, *Raids near Hanoi,* p. 1; The New York Times, *U.S. reports bombing of a power plant,*

p. 15

[27] Ibid; C. Whitney, *Air strikes close to center of Hanoi,* p. 1

[28] R. Francis, Author Interview, Transcript 1

[29] Ibid

[30] Ibid; F-4E Manual

[31] R. Francis, Author Interview, Transcript 1

[32] Ibid

[33] Ibid; Veteran Tributes Bio, *Richard Francis*;

[34] P.O.W. Network Bio, *Farrell Sullivan*; Wall of Faces, *Farrell Junior Sullivan*

[35] J. Hubbell, *P.O.W.*, p. 605-618; M. Porter, *The First 120 Days,* p. 71; C. Hobson, *Vietnam Air Losses,* p. 234-230

[36] J. Morrocco, *Air War 1969-1973*, p. 136; J. Sherwood, *Nixon's Trident,* p. 64; E. Hartsook and S. Slade, *Air War,* p. 299. 301; J. Smith, *Linebacker Raids,* p. 83; W. Thompson, *To Hanoi and Back,* p. 248

[37] D. Whitfield, *Historical and Cultural Dictionary of Vietnam*, p. 42; CIA, *Dike Bombing Issue,* p. 5-7; W. Parks, *Linebacker and the Law of War*; P. Asselin, *A Bitter Peace,* p. 60

[38] The New York Times, *Report on Bombing of Dikes*, p. 2; W. Parks, *Linebacker and the Law of War*

[39] P. Asselin, *A Bitter Peace,* p. 58-60

[40] C. Whitney, *Hanoi says raids struck at dikes,* p. 1

[41] Time Magazine, *Battle of the Dikes*

[42] K. Smith, *Jane Fonda's Vietnam Actions*

[43] E. Meredith, *Jane Fonda,* p. 3

[44] E. Meredith, *Jane Fonda,* p. 5-10; *Further reports on Jane Fonda's activities in DRV*; *Jane Fonda Statement to US Pilots*

[45] Time Magazine, *Battle of the Dikes*

[46] Ibid; CIA, *Possible alternatives to Rolling Thunder,* p. 2;

[47] W. Parks, *Linebacker and the Law of War;* The New York Times, *Intelligence report on bombing dikes*, p. 2; CIA, *Dike Bombing Issue*

[48] C. Honour, Interview Transcript 3; Command History, *Saratoga*, 1972, p. 4

[49] C. Honour, Interview Transcript 3; B. Jellison, *RA-5C Vigiliante*

[50] Ibid

[51] Ibid

[52] S. Hersh, War Foes, p. 3

[53] The New York Times, *Intelligence report on bombing dikes*, p. 2; CIA, *Dike Bombing Issue*; Time Magazine, *Battle of the Dikes*

[54] W. Parks, *Linebacker and the Law of War*

[55] M. Michel, *Clashes,* p. 248

[56] J. Correll, *Wild Weasels in Vietnam*; Society of Wild Weasels, *We Remember*

[57] E. Eiler, Author Interview, Transcript; Vietnam War Commemoration, *Fight to*

Save F-105 Aviators

[58] Ibid

[59] M. Michel, *Clashes,* p. 248

[60] C. Hobson, *Vietnam Air Losses,* p. 224-230

[61] M. Michel, *Clashes,* p. 244; C. Hobson, *Vietnam Air Losses,* p. 271; D. McCarthy, *MiG Killers,* p. 156

[62] W. Thompson, *To Hanoi and Back,* p. 240; D. McCarthy, *MiG Killers,* p. 155- 157

[63] Ibid

[64] J. Morrocco, *Air War,* p. 144

[65] D. Leone, *Loose Deuce vs Fluid Four;* M. Michel, *Clashes,* p. 284

[66] Ibid

[67] MACV, *1972-1973 Command History*, p. 67-68

[68] S. Emerson, *Vietnam's Final Air Campaign,* p. 76

Chapter 6

[1] N. Truong, *The Easter Offensive of 1972,* p. 128; D. Andrade, *America's Last Vietnam Battle,* p. 430

[2] D. Andrade, *America's Last Vietnam Battle,* p. 430, 434; N. Truong, *The Easter Offensive of 1972,* p. 130

[3] Ibid

[4] C. Melson and C. Arnold, *Marines in Vietnam,* p. 160-161

[5] S. Massimini, Author Interview, Transcript 1; C. Schuster, *Terror in the Skies*

[6] S. Massimini, Author Interview, Transcript 1

[7] C. Melson and C. Arnold, *Marines in Vietnam,* p. 162-164, 214

[8] S. Massimini, Author Interview, Transcript 1

[9] J. Willbanks, *The Battle of An Loc*, p. 56-57

[10] J. Willbanks, *The Battle of An Loc*, p. 57-58

[11] J. Willbanks, *The Battle of An Loc*, p. 60

[12] J. Willbanks, *The Battle of An Loc*, p. 60-70; C. Melson and C. Arnold, *Marines in Vietnam,* p 161

[13] N. Truong, *The Easter Offensive of 1972,* p. 169-171

[14] D. Andrade, *America's Last Vietnam Battle,* p. 282; N. Truong, *The Easter Offensive of 1972,* p. 91-92

[15] N. Truong, *The Easter Offensive of 1972,* p. 95-96

[16] D. Andrade, *America's Last Vietnam Battle,* p. 282; N. Truong, *The Easter Offensive of 1972,* p. 97; AMCOM, *TOW*

[17] N. Truong, *The Easter Offensive of 1972,* p. 98

[18] D. Fulghum & T. Maitland, *The Vietnam Experience*, p. 157; L. Sorley, *A Better War,* p. 344

[19] D. Andrade, *America's Last Vietnam Battle,* p. 299

[20] Ibid

[21] D. Andrade, *America's Last Vietnam Battle,* p. 329; N. Truong, *The Easter Offensive of 1972,* p. 101

[22] Ibid

[23] D. Andrade, *America's Last Vietnam Battle,* p. 229, 325

[24] D. Andrade, *America's Last Vietnam Battle,* p. 299, 329; N. Truong, *The Easter Offensive of 1972,* p. 171

[25] N. Truong, *The Easter Offensive of 1972,* p. 104-105

[26] Ibid

[27] N. Truong, *The Easter Offensive of 1972,* p. 54

[28] D. Andrade, *America's Last Vietnam Battle,* p. 161, 170; C. Hobson, *Vietnam Air Losses,* p. 223-227

[29] D. Andrade, *America's Last Vietnam Battle,* p. 168, 170

[30] D. Andrade, *America's Last Vietnam Battle,* p. 167-170

[31] N. Truong, *The Easter Offensive of 1972,* p. 62

[32] N. Truong, *The Easter Offensive of 1972,* p. 65-66

[33] N. Truong, *The Easter Offensive of 1972,* p. 66; D. Andrade, *America's Last Vietnam Battle,* p. 187

[34] N. Truong, *The Easter Offensive of 1972,* p. 66; P.O.W. Network, *Hendrix, Jerry*

[35] N. Truong, *The Easter Offensive of 1972,* p. 67-68; D. Fulgham and T. Maitland,

p. 181

[36] N. Truong, *The Easter Offensive of 1972,* p. 67; D. Andrade, *America's Last Vietnam Battle,* p. 181, 191

[37] D. Andrade, *America's Last Vietnam Battle,* p. 191-192, 327

[38] N. Truong, *The Easter Offensive of 1972,* p. 70

[39] Ibid

[40] N. Truong, *The Easter Offensive of 1972,* p. 70-71

[41] Ibid

[42] N. Truong, *The Easter Offensive of 1972,* p. 71, 74

[43] D. Andrade, *Trial by Fire,* p. 536

[44] D. Andrade, *Trial by Fire,,* p. 529

[45] Naval History and Heritage Command. *H-Gram 074*

[46] D. Andrade, *America's Last Vietnam Battle,* p. 487. Author Andrade notes that unofficial South Vietnamese casualties figures may have been twice this number.

[47] N. Truong, *The Easter Offensive of 1972,* p. 177

[48] P. Asselin, *A Bitter Peace,* p. 59

Chapter 7

[1] J. Bowman, *Almanac of Vietnam War,* p. 318;

[2] M. Michel, *Clashes,* p. 250

[3] J. Smith, *The Linebacker Raids,* p. 103; W. Thompson, *To Hanoi and Back,* p. 92

[4] National Museum of the United States Air Force, *F-111A Aardvark*; C. Hobson, *Vietnam Air Losses,* p. 237

[5] C. Hobson, *Vietnam Air Losses,* p. 237; J. Smith, *The Linebacker Raids,* p.108; W. Thompson, *To Hanoi and Back,* p. 246

[6] *Tragedy Aboard the USS Stoddert*

[7] Naval History & Heritage Command, *H-Gram 074*

[8] Ibid

[9] Ibid

[10] Ibid

[11] Ibid; Naval History & Heritage Command, *The Human Cost*

[12] The New York Times, *North Vietnamese are now requiring everyone to work,* p. 3

[13] C. Hobson, *Vietnam Air Losses,* p. 235-238

[14] M. Browne, *U.S. Planes Raid 4 Bases,* p. 1; J. Bowman, *Almanac,* p.

[15] E. Tilford, *Setup,* p. 237

[16] J. Smith, *The Linebacker Raids,* p. 115; W. Thompson, *To Hanoi and Back,* p. 242

[17] J. Smith, *The Linebacker Raids,* p. 110; R. Boniface, *MiGs over North Vietnam,* p. 118

[18] M. Faram, *Race riot at sea*

[19] The New York Times, *U.S. Copter Fire, p. 3*

[20] B. Gwertzman, *France's Mission in Hanoi Wrecked,* p. 1; J. Bowman, *Almanac,* p. 324-325; The New York Times, *Hanoi Area Raids Reported Curbed,* p. 11; [21] The New York Times, *2nd Biggest Bombing,* p. 77; R. Futrell et al, *Aces and Aerial Victories,* p. 109-111

[22] J. Willbanks, *Vietnam War Almanac*, p. 421

[23] P. Asselin, *A Bitter Peace,* p. 76-77

[24] L. Nguyen, *Hanoi's War,* p. 260

[25] Ibid; D. Schmitz, *Richard Nixon,* p. 140; P. Asselin, *A Bitter Peace,* p. 180

[26] D. Schmitz, *Richard Nixon,* p. 140

[27] Ibid

Chapter 8

[1] P. Asselin, *A Bitter Peace,* p. 55; Vietnam, *Interview with John Negroponte*; R. Reeves, *President Nixon,,* p. 474

[2] K. Hughes, *Fatal Politics,* p. 9; Presidential Recordings, *Nixon and Kissinger*

[3] Presidential Recordings, *Nixon and Kissinger*

[4] P. Asselin, *A Bitter Peace,* p. 55

[5] P. Asselin, *A Bitter Peace,* p. 69

[6] P. Asselin, *A Bitter Peace,* p. 57-58

[7] R. Reeves, *President Nixon,,* p. 523, 531

[8] P. Asselin, *A Bitter Peace,* p. 65-66

[9] J. White, *Civil Affairs in Vietnam*, p. 10; P. Asselin, *A Bitter Peace,* p. 66

[10] P. Asselin, *A Bitter Peace,* p. 66

[11] P. Asselin, *A Bitter Peace,* p. 66-67, 72

[12] P. Asselin, *A Bitter Peace,* p. 69-70; D. Schmitz, *Richard Nixon,* p. 140

[13] P. Asselin, *A Bitter Peace,* p. 70

[14] R. Reeves, *President Nixon,,* p. 527

[15] P. Asselin, *A Bitter Peace,* p. 72; R. Reeves, *President Nixon,,* p. 527-528

[16] P. Asselin, *A Bitter Peace,* p. 74

[17] R. Reeves, *President Nixon,,* p. 528

[18] P. Asselin, *A Bitter Peace,* p. 76

[19] R. Reeves, *President Nixon,,* p. 528-529

[20] P. Asselin, *A Bitter Peace,* p. 76

[21] Office of the Historian, *21. Memorandum of Conversation, Paris October 8*

[22] P. Asselin, *A Bitter Peace,* p. 79

[23] Ibid; Office of the Historian, *21. Memorandum of Conversation, Paris October 8*

[24] R. Reeves, *President Nixon,,* p. 529

[25] R. Reeves, *President Nixon,,* p. 530; P. Asselin, *A Bitter Peace,* p. 76

[26] P. Asselin, *A Bitter Peace,* p. 80; Office of the Historian, *21. Memorandum of Conversation, Paris October 8;* R. Reeves, *President Nixon,,* p. 529-530

[27] R. Reeves, *President Nixon,,* p. 530; Office of the Historian, *23. Memorandum of Conversation, Paris October 10*

[28] P. Asselin, *A Bitter Peace,* p. 81

[29] Office of the Historian, *24. Memorandum of Conversation, Paris October 11-12;*

R. Reeves, *President Nixon,,* p. 533

[30] P. Asselin, *A Bitter Peace,* p. 81-82

[31] P. Asselin, *A Bitter Peace,* p. 82-85

[32] Ibid

[33] Office of the Historian, *24. Memorandum of Conversation, Paris October 11-12;*

S. Emerson, *Vietnam's Final Air Campaign,* p. 86; R. Nixon, *Memoirs of Richard Nixon,* p. 693-694

[34] P. Asselin, *A Bitter Peace,* p. 86

[35] R. Reeves, *President Nixon,,* p. 534-535; Office of the Historian, *Memo from Laird to Nixon*

[36] R. Reeves, *President Nixon,,* p. 534-535

[37] P. Asselin, *A Bitter Peace,* p. 99-100

[38] R. Reeves, *President Nixon,,* p. 535

[39] R. Reeves, *President Nixon,,* p. 535; T. Van Don, *Our Endless War*, p. 202

[40] P. Asselin, *A Bitter Peace,* p. 90

[41] P. Asselin, *A Bitter Peace,* p. 90-91

[42] Ibid; R. Reeves, *President Nixon,,* p. 536-537

[43] P. Asselin, *A Bitter Peace,* p. 98-99

[44] P. Asselin, *A Bitter Peace,* p. 92

[45] P. Asselin, *A Bitter Peace,* p. 93

[46] P. Asselin, *A Bitter Peace,* p. 93-94

[47] R. Reeves, *President Nixon,,* p. 536

[48] H. Kissinger, *White House Years*, p. 1375-1376

[49] P. Asselin, *A Bitter Peace,* p. 94-95

[50] R. Reeves, *President Nixon,,* p. 537

[51] R. Reeves, *President Nixon,,* p. 537-538

[52] P. Asselin, *A Bitter Peace,* p. 96-97

[53] P. Asselin, *A Bitter Peace,* p. 97-98

[54] P. Asselin, *A Bitter Peace,* p. 101, 111

[55] P. Asselin, *A Bitter Peace,* p. 98

[56] J. Smith, *The Linebacker Raids,* p. 115; C. Hobson, *Vietnam Air Losses,* p. 271;

J. Hubbell, *P.O.W.*, p. 606-618

[57] R. Reeves, *President Nixon,,* p. 538-539; P. Asselin, *A Bitter Peace,* p. 111

[58] P. Asselin, *A Bitter Peace,* p. 101; S. Emerson, *Vietnam's Final Air Campaign,* p. 87; J. Bowman, *Almanac of Vietnam War,* p. 327

[59] Office of the Historian. *73. Editorial Note*

[60] R. Reeves, *President Nixon,,* p. 540

[61] P. Asselin, *A Bitter Peace,* p. 101

[62] H. Kissinger, *White House Years,* p. 1400

[63] R. Nixon, *Memoirs of Richard Nixon,* p. 705

[64] A. Haig, *Inner Circles*, p. 302

[65] R. Reeves, *President Nixon,,* p. 541

[66] P. Asselin, *A Bitter Peace,* p. 111-112

[67] Office of the Historian (OOTH), *26. Memorandum of Conversation: Paris, November 20; 27. Memorandum of Conversation: Paris, November 21, 1972*

[68] P. Asselin, *A Bitter Peace,* p. 110

[69] P. Asselin, *A Bitter Peace,* p. 121-122

[70] OOTH, *32. Memorandum of Conversation: Paris, December 4*

[71] OOTH, *33. Memorandum of Conversation: Paris, December 4*

[72] Ibid

[73] OOTH, *34. Memorandum of Conversation: Paris, December* 6

[74] Ibid; P. Asselin, *A Bitter Peace,* p. 129

[75] OOTH, *35. Memorandum of Conversation: Paris, December 7*

[76] OOTH, *36, Memorandum of Conservation: Paris, December 8*

[77] OOTH, *38. Memorandum of Conversation: Paris, December 9*

[78] Ibid

[79] OOTH, *40. Memorandum of Conversation: Paris, December 12*

[80] P. Asselin, *A Bitter Peace,* p. 138

[81] Ibid

[82] Ibid

[83] A. Haig, *Inner Circles,* p. 309

[84] OOTH, *41. Memorandum of Conversation: Paris, December 13*

[85] A. Haig, *Inner Circles,* p. 309; OOTH, *189. Nixon Letter to Thieu, December 17*

[86] A. Haig, *Inner Circles,* p. 310-311

[87] P. Asselin, *A Bitter Peace,* p. 141-142

[88] OOTH, *206. Backchannel Message, Haig to Kissinger* (including Thieu Letter to Nixon)

[89] Ibid

[90] P. Asselin, *A Bitter Peace,* p. 148-149

[91] H. Kissinger, *White House Years*, p. 1467

[92] R. Nixon, *Memoirs of Richard Nixon,* p. 734

[93] H. Kissinger, *White House Years*, p. 1447-1448

[94] P. Asselin, *A Bitter Peace,* p. 144

[95] J. Morrocco, *Rain of Fire*, p. 146

[96] K. Eschmann, *Linebacker*, p. 75

[97] K. Eschmann, *Linebacker*, p. 74

[98] P. Asselin, *A Bitter Peace,* p. 144

[99] S. Emerson, *Vietnam's Final Air Campaign,* p. 93

[100] J. Smith, *The Linebacker Raids,* p. 121; S. Emerson, *Vietnam's Final Air Campaign,* p. 90, 93

[101] S. Williams, Author Interview, Transcript 1

[102] Ibid

[103] K. Eschmann, *Linebacker*, p. 77; J. Smith, *The Linebacker Raids,* p. 121; S. Emerson, *Vietnam's Final Air Campaign,* p. 93

[104] K. Eschmann, *Linebacker*, p. 74

[105] G. Porter, *Peace Denied*, p. 365

Chapter 9

[1] McCarthy and Allison, *A View from the Rock*, p. 48

[2] R. Certain, *Unchained Eagle,* location (l). 484

[3] S. Emerson, *Vietnam's Final Air Campaign,* p. 93

[4] Ibid; R. Certain, *Unchained Eagle,* l. 324

[5] McCarthy and Allison, *A View from the Rock*, p. 42

[6] R. Certain, Author Interview, Transcript 1

[7] M. Pribbenow (translator), *Victory in Vietnam,* p. 316-319; P. Asselin, *A Bitter Peace,* p. 122

[8] R. Certain, *Unchained Eagle,* l. 596; W. Parks, *Linebacker and the Law of War*

[9] K. Eschmann, *Linebacker*, p. 91; McCarthy and Allison, *A View from the Rock*, p. 42

[10] R. Certain, *Unchained Eagle,* l. 623

[11] W. Boyne, *Linebacker II*, p. 55; R. Certain, Author Interview, Transcript 1

[12] R. Certain, Author Interview, Transcript 1; R. Certain, *Unchained Eagle,* l. 651, 659

[13] R. Certain, Author Interview, Transcript 1; R. Certain, *Unchained Eagle,* l. 680, 688

[14] Ibid; R. Certain, *Unchained Eagle,* l. 718, 729, 741; C. Johnson, *Linebacker Operations,* p. 31

[15] Veteran Tributes, *Walter Ferguson, Donald Rissi, Robert Thomas*

[16] C. Hobson, *Vietnam Air Losses,* p. 242; McCarthy and Allison, *A View from the Rock*, p. 59-61;

[17] J. Smith, *The Linebacker Raids,* p. 129; K. Eschmann, *Linebacker*, p. 107; E. Hartsook and S. Slade, *Air War,* p. 333

[18] Ibid

[19] J. Tayabji, *Christmas Bombing*

[20] Ibid, Hartsook and Slade; G. Joiner, A. Dean, *Linebacker II Retrospective,* p. 17

[21] M. Gonzalez, *Forgotten History,* p.4; C. Hobson, *Vietnam Air Losses,* p. 245; A. Kohloff, *BUFs Go Downtown,* p. 50

[22] C. Hobson, *Vietnam Air Losses,* p. 242-243; J. Sherwood, *Nixon's Trident,* p. 71-72

[23] K. Eschmann, *Linebacker*, p. 111, 139-140

[24] K. Eschmann, *Linebacker*, p. 112-115; McCarthy and Allison, *A View from the Rock*, p. 68;

G. Joiner, A. Dean, *Linebacker II Retrospective,* p. 18-19

[25] K. Eschmann, *Linebacker*, p. 116; C. Hobson, *Vietnam Air Losses,* p. 242-243

[26] K. Eschmann, *Linebacker*, p. 131-133; C. Hobson, *Vietnam Air Losses,* p. 243

[27] G. Joiner, A. Dean, *Linebacker II Retrospective,* p. 22; McCarthy and Allison, *A View from the Rock*, p. 91

[28] G. Joiner, A. Dean, *Linebacker II Retrospective,* p. 23

[29] K. Eschmann, *Linebacker*, p. 137, 140

[30] C. Hobson, *Vietnam Air Losses,* p. 244; K. Eschmann, *Linebacker*, p. 147; G. Joiner, A. Dean, *Linebacker II Retrospective,* p. 23-24; J. Morrocco, *Air War 1969-1973*, p. 160

[31] G. Joiner, A. Dean, *Linebacker II Retrospective,* p. 26

[32] Ibid

[33] C. Hobson, *Vietnam Air Losses,* p. 244

[34] K. Eschmann, *Linebacker*, p. 154; G. Joiner, A. Dean, *Linebacker II Retrospective,* p. 26

[35] McCarthy and Allison, *A View from the Rock*, p. 97

[36] S. Emerson, *Vietnam's Final Air Campaign,* p. 106-107

[37] K. Eschmann, *Linebacker*, p. 156, 159; McCarthy and Allison, *A View from the*

Rock, p. 103

[38] McCarthy and Allison, *A View from the Rock*, p. 112; D. Branum, *MiG Kill by 'Diamond Lil'*

[39] McCarthy and Allison, *A View from the Rock*, p. 107

[40] K. Eschmann, *Linebacker*, p. 163

[41] C. Johnson, *Linebacker Operations,* p. 100-101

[42] K. Eschmann, *Linebacker*, p. 163; McCarthy and Allison, *A View from the Rock*, p. 119

[43] K. Eschmann, *Linebacker*, p. 150; McCarthy and Allison, *A View from the Rock*, p. 140

[44] McCarthy and Allison, *A View from the Rock*, p. 136-137

[45] S. Bach, *Kham Thien Bombing*; P. Asselin, *Bitter Peace,* p. 150

[46] P. Asselin, *Bitter Peace,* p. 150-151

[47] McCarthy and Allison, *A View from the Rock*, p. 156

[48] K. Eschmann, *Linebacker*, p. 183; McCarthy and Allison, *A View from the Rock*, p. 152

[49] C. Hobson, *Vietnam Air Losses,* p. 246

[50] K. Eschmann, *Linebacker*, p. 183, 190-192

[51] R. Bernier, *The MiG Hunters*; D. McCarthy, *MiG Killers*, p. 157

[52] McCarthy and Allison, *A View from the Rock*, p. 168, 174-175; C. Johnson, *Linebacker Operations,* p. 95

[53] S. Emerson, *Vietnam's Final Air Campaign,* p. 90-91, 93; R. Futrell et al, *Aces and Aerial Victories,* p. 16-17

[54] B. Nalty, *Air War,* p. 182; C. Johnson, *Linebacker Operations,* p. 95

[55] McCarthy and Allison, *A View from the Rock*, p. 173; P. Asselin, *A Bitter Peace,* p. 156

[56] P. Asselin, *A Bitter Peace,* p. 156-157, 180

[57] K. Eschmann, *Linebacker*, p. 57

[58] M. Pribbenow (translator), *Victory in Vietnam,* p. 328

[59] Ibid; P. Asselin, *A Bitter Peace,* p. 157

[60] A. Haig, *Inner Circles,* p. 313; Thompson and Frizzell, *Lessons,* p.105. For more on Thompson's thinking, see also p. 143, 168-170, and 177

[61] P. Asselin, *A Bitter Peace,* p. 155-156, p. 152-153; D. Rosenbaum, *Efficacy of the Bombing,* p. 10

[62] McCarthy and Allison, *A View from the Rock*, p. 176-177

Chapter 10

[1] OOTH, *256. Cable from Kissinger to Nixon,* January 9; *The Vietnam Agreement and Protocols*

[2] Ibid, OOTH, *256.*

[3] P. Asselin, *A Bitter Peace,* p. 161; *The Vietnam Agreement and Protocols*

[4] OOTH, *254. Letter from Thieu to Nixon,* January 7

[5] P. Asselin, *A Bitter Peace,* p. 161-163; *The Vietnam Agreement and Protocols*

[6] Ibid, *The Vietnam Agreement and Protocols*

[7] Ibid

[8] OOTH, *45. Memorandum of Conversation,* January 11

[9] OOTH, *290. Letter from Nixon to Thieu,* January 17

[10] OOTH, *310. Letter from Thieu to Nixon,* January 20

[11] OOTH, *313. Letter from Nixon to Thieu,* January 20

[12] P. Asselin, *A Bitter Peace,* p. 173-174

[13] OOTH, *313. Letter from Nixon to Thieu,* January 20

[15] H. Kissinger, *White House Years,* p. 1470; R. Nixon, *Memoirs of Richard Nixon,* p. 751; R. Reeves, *President Nixon,* p. 562-563

[16] OOTH, *325. Letter from Nixon to Thieu,* January 22

[17] P. Asselin, *A Bitter Peace,* p. 174-175, 181; *The Vietnam Agreement and Protocols*

[18] T. DePastino, *Nixon Announces*

[19] Ibid

[20] F. Lewis, *Vietnam Peace Pacts Signed*; OOTH, *341. Telegram from the U.S. Delegation,* 27 January

[21] P. Asselin, *A Bitter Peace,* p. 177-178

Chapter 11

[1] Pacific Airlifter, *Loc Ninh Prisoner Exchange*

[2] Alvarez & Pitch, *Chained Eagle: The Heroic Story of the First American Shot Down over North Vietnam,* 2005; Everett Alvarez, interview transcript 1, conducted by J. Keith Saliba on 12/24/21; Sterba, "Airlift is Begun." Special to the New York Times, February 12, 1973.

[3] Alvarez, & Pitch, *Chained Eagle;* Alvarez, interview; Philpott, *Glory Denied: The Saga of Vietnam Veteran Jim Thompson, America's Longest-Held Prisoner of War,* 2002.

[4] The Long Road Home: U.S. Prisoner of War Policy and Planning in Southeast Asia, Vernon E. Davis, Historical Office, Office of the Secretary of Defense, 2000,

p. 491.

[5] Ibid

[6] Operation Homecoming: MAC's Finest Hour, Coy F. Cross II, 9th Reconnaissance Wing, History Office, 1999; Davis, The Long Road Home, 492- 493.

[7] Alvarez, & Pitch, *Chained Eagle*; Alvarez, interview, J. Keith Saliba; Vernon, The Long Road Home, pp. 499-500: Cross, MAC's Finest Hour.

[8] Alvarez, & Pitch, *Chained Eagle*; History of Operation Homecoming, 6 September 1971 to 27 July 1973, Headquarters, Thirteenth Air Force, Clark Air Base, R.P., p. I-17.

[9] Alvarez, & Pitch, *Chained Eagle*, p. 257; Cross, MAC's Finest Hour, pp. 30-31.

[10] Operation Homecoming: Inception through Implementation, 1964-1973, Air Force Historical Research Association, released 05/08/1984; Cross, MAC's Finest Hour.

[11] Alvarez, & Pitch, *Chained Eagle*, p. 257.

[12] Mikeline "Mickey" Mantel interview conducted by J. Keith Saliba on 1/28/22, transcript 1

[13] Cross, MAC's Finest Hour; Risner, *The Passing of the Night,* pp. 246-247.

[14] Alvarez, & Pitch, *Chained Eagle*; Cross, MAC's Finest Hour.

[15] Alvarez & Pitch, *Chained Eagle,* p. 259.

[16] Risner, *The Passing of the Night,* 249.

[17] Denton, *When Hell w as in Session,* p. 235.

[18] Alvarez & Pitch, *Chained Eagle,* p. 271.

[19] OOTH, *6. Address by President Nixon, March 29, 1973*